MW01519201

SIR HUGH MYDDELTON.

[*Frontispiece.*

LIVES OF THE ENGINEERS

VERMUYDEN—MYDDELTON—PERRY— JAMES BRINDLEY

BY SAMUEL SMILES

AUTHOR OF 'CHARACTER,' 'SELF-HELP,' ETC.

EARLY ENGINEERING

" Bid Harbours open, Public Ways extend;
Bid Temples, worthier of God, ascend ;
Bid the broad Arch the dang'rous flood contain,
The Mole projected break the roaring main,
Back to his bounds their subject sea command,
And roll obedient rivers through the land.
These honours, Peace to happy Britain brings :
These are imperial works, and worthy kings."

POPE

POPULAR EDITION

LONDON
JOHN MURRAY ALBEMARLE STREET
1904

PRINTED BY
WILLIAM CLOWES AND SONS, LIMITED,
LONDON AND BECCLES.

INTRODUCTION.

Lord Bacon hath truly said that there are three things which make a nation great and prosperous, —a fertile soil, busy workshops, and easy conveyance for men and commodities from one place to another.

The soil of England is sufficiently fertile; but it has been made so principally by the industry of its people. A large portion of it, and that the most fertile, has been reclaimed and embanked from the sea. The two thousand square miles of land in Lincolnshire and Cambridgeshire, forming the Great Level of the Fens, would, but for human skill and industry, be lying under water, the haunt of wild birds.

Who does not know of the business of our Workshops? — of the employment they give to large numbers of the population; of the tools they fabricate; of the articles of commerce they produce; of the demands they supply for consumers in all parts of the world.

Then, with respect to Lord Bacon's " easy conveyance for men and commodities from one place to another:" have we not our Roads, our Canals, and our Railways; and has not England also been the originator of the Steam engine, the Steamboat, and the Screw ship?

Although the inventions of British engineers now exercise so great an influence all over the world, it may possibly excite some surprise to find how very recently England has taken up her present position in regard to Engineering. Indeed, skilled Industry may be said to form the youngest outgrowth of our national life.

Most of the continental nations had a long start of us in art, in science, in mechanics, in navigation, and in engineering. In former times we were more given to fighting than working. The wars between England and France—between England and Scotland—between England and Ireland—deterred our people from engaging in the pursuits of industry and commerce. The only article we exported was wool. We seem to have been unable to make our own clothing; for a large part of the exported wool was returned to England as cloth, for sale in the English markets. We grew wool for Flanders, as America grows cotton for England now. Even the small quantity of cloth manufactured at home was sent to the Low Countries to be dyed.

Many branches of industry have been established in England by foreigners. The Flemings introduced cloth-making, stuff-making, and linen-weaving. The brothers Elers, Dutchmen, began the pottery manufacture. Spillman, a German, established the first paper-mill at Dartford; and Bromen, another Dutchman, brought the first coach into England. Towards the end of the seventeenth century, a large number of Huguenot manufacturers established various branches of industry in England and Ireland which, until then, had for the most part been conducted abroad.

We owe the first working of our mines to the Germans; who brought with them nearly all the mining terms still in use among us. Elizabeth invited skilled miners from Germany to settle in different parts of England, for the purpose of teaching our people the best methods of working the ores. Two of these, Hochstetter and Thurland, of Augsburg, established copper-works at Keswick, which were worked to such advantage that the Queen said she had left more brass than she had found iron ordnance in England.

Not many centuries since, Italy, Spain, France, and Holland, looked down contemptuously on the poor but proud islanders, contending with nature for a sustenance amidst their fogs and their mists. Though surrounded by the sea, we had scarcely any Royal navy until within the last three hundred years. An old writer, in his eulogium of Henry VII., said: "He built a chapel; he built a ship." But we had no Commercial navy. The carrying trade of the world was then performed by the Dutch, the Genoese, the Venetians, and the Portuguese.

When we wanted any skilled work done, we almost invariably sent abroad for foreigners to do it. Our first ships were built by Danes, Venetians, and Genoese. When the royal ship, the 'Mary Rose,' sank at Spithead in 1545, a company of Venetians was hired to raise her. On that occasion Peeter de Andreas was employed, assisted by his ship carpenters and sailors, with "sixty English maryners to attend upon them."

Our first lessons in mechanical and civil engineering were principally obtained from the Flemings and Dutchmen, who supplied us with our first wind-mills, water-mills, and fulling-mills. When an engine was required to pump water from the Thames, Peter Morice, the Dutchman, was employed to erect it. The first iron cannons were cast at Buxted, in Sussex, by Peter Baude, a Frenchman; and the first mortar pieces by Peter Van Collet, a Flemish gunsmith.

Holland even sent us the necessary engineers and labourers to execute our first great works of drainage. The Great Level of the Fens was drained by Vermuyden, a native of Zealand. Another Dutchman, Freestone, was employed to reclaim the marsh, near Wells, in Norfolk. Canney Island, near the mouth of the Thames, was embanked by Joas Croppenburgh and his company of Dutch workmen.

When a new haven was required at Yarmouth, Joas Johnson, the Dutch engineer, was employed to plan and construct the works. When Dover harbour was threatened with ruin in the reign of Henry VIII., Ferdinand Poins, a Fleming, was invited to come over to plan a new quay for the purpose of preserving the haven. Matthew Hake, of Gravelines, in Flanders, was also sent for about the same time, for the purpose of repairing a serious breach which had been made in the banks of the river Witham, at Boston; and he brought with him not only the mechanics but the manufactured iron required for the work.

Indeed very little English iron was then used. Almost the only iron in England was imported by the Foreign Merchants of the Steelyard Company. The Spaniards prided themselves upon the superiority of their armour, and regarded the scarcity of iron in England as an important element of success in their invasion of this country by the great Armada. Down to the end of last century, England was principally indebted for its iron to Sweden, Germany, Spain, and Russia.

Yet, a very long time ago, there must have been great architects as well as great builders in England. This is sufficiently proved by the magnificent cathedrals, abbeys, and ancient bridges which still exist. It is not improbable that these old architects and builders were Frenchmen, the western part of France being for many hundred years subject to the English crown. However that may be, the English architects and builders must have greatly fallen off by the time of Elizabeth. Thus, when Sir Thomas Gresham built the Royal Exchange in 1566, he brought from Flanders the requisite masons and carpenters to execute it under the direction of Henryke, their master builder. The foreigners also brought with them all necessary materials— the wainscot, the glass, the slates, the iron, and even much of the stone for the building. In short, as Holinshed relates, Gresham "bargained for the whole mould and substance of his workmanship in Flanders." The only Englishmen employed upon the work, were the labourers who carried the lime and stone required for the building.

Bridge building also must have greatly fallen off. Little more than a hundred years ago, there were few architects or masons able to build a bridge of any extent; and when a second bridge had to be erected over the Thames at London, the French engineer Labelye, a native of Switzerland, was sent for to build it; though the original Westminster Bridge is already destroyed, and replaced by a new structure.

In short, we depended for our engineering, even

b 2

more than we did for our pictures or our music,
upon foreigners. At a time when Holland had
completed its magnificent system of water communication, and when France, Germany, and even
Russia, had opened up important lines of inland
communication, England had not cut a single canal,
while our roads were about the worst in Europe.

The first English canal was made by a workman
taken from the lowest rank of society. He could
not write his own name. But he was a persevering
ingenious man, possessed of a shrewd mother
wit. Brindley would have become great in his own
trade—that of a millwright; but having succeeded
in accomplishing several important works which
he had undertaken, he attracted the attention of
the Duke of Bridgewater, who employed him to
make the first English canal between Worsley and
Manchester.

Water was the great difficulty of the early Engineer. First he had to dam it out or bale it out, like
Vermuyden. Or he erected a lighthouse upon a
rock in the ocean, to withstand its power, like
Smeaton. Or he founded piers, bridges, and docks,
in the midst of it, like Rennie. The Engineer had
also to make water serve his purposes. He led it
into a mill-stream, and made it grind his corn and
work his machinery. Or he collected it into reservoirs to supply a large city with water, like Myddelton; or, like Brindley, he made it fill long canals,
extending between one town and another, and thus
made a great water road to convey coal, and
minerals, and merchandise. Thus, in the hands
of the Engineer, water, instead of being a tyrant,
became a servant; instead of being a destroyer, it
became a useful labourer and a general civiliser.

Richard Cobden has said, that "the opening up of the internal communications of a country, is undoubtedly the first and most important element of its growth in commerce and civilisation. Hence our canals were regarded by Baron Dupin as the primary material agents of the wealth of Great Britain." Roads, however, are more important than canals; and at the time when the first canals were made, the roads of England still remained in a deplorable condition.

Down to the middle of the last century, it was almost impossible to travel with speed or comfort in any direction. Everything was carried on horse-back, and sometimes on bullock-back. Corn, coal, wool, iron, and such like articles, were carried through the country on pack-horses. The roads were not roads so much as tracks; and when the old tracks became dangerous through depth of mud, new tracks were struck out across the adjoining fields. Guides were employed to keep travellers out of the mud; for it was often thick and slab. During the civil war, some eight hundred horse were taken prisoners while sticking in the mud in Buckinghamshire.

When all merchandise was carried by pack-horses over bad roads, domestic commerce was simply impossible. Hence the commerce of London was mostly Foreign. We bought cloth from Belgium, silk from France, cutlery from Italy, hats from Flanders, Delftware from Holland, and iron from Spain, Sweden, and Russia. It cost so much less to bring goods from Hamburg, Amsterdam, or Havre by sea, than from Norwich, Birmingham, or Manchester by land.

The English highroads were also a good deal

beset by highwaymen. The traveller often passed along bridle ways through fields, where frequent gibbets warned him of his perils.

When the roads were attempted to be made suitable for carts, waggons, and carriages, they were very little improved. The stone or metal was put on, and the roads were simply left to manage themselves. They soon became mere slumpers of stones and mud. If they were mended at all, it was only by throwing big stones into the biggest holes. Arthur Young when travelling in the northern counties of England, in 1768, measured the depth of the tracks he travelled through. Between "proud Preston" and Wigan, he says, "I actually measured ruts of four feet deep, floating with mud only from a wet summer; and between the towns I actually passed three carts broken down in those eighteen miles of execrable memory."

In another place, he says, "You will form a clear idea of these roads, if you suppose them to represent the roofs of houses, and the roads to run across them." In winter, he says, it would have cost no more money to make the roads *navigable*, than to make them hard. Even near London, the roads were often impassable gulfs of mud. It then took a longer time for a carriage to reach twenty miles out of London, than it now does to reach York or Manchester.

An Act was passed in 1663, authorising the first turn-pike roads to be made, with the usual toll-gates ; but the people would not bear the tolls, and the gates were pulled down wherever erected. As late as 1753, a general raid was made upon the toll-gates in Yorkshire. About a dozen were destroyed weekly. The soldiers were at length sent to protect the turnpikes; and on their apprehending a score of the rioters to convey them to York, the people attempted a rescue, when the soldiers fired, and killed and wounded a considerable number.

Nevertheless, considerable progress was made in the improvement of the roads, especially after the last Highland rebellion. Between 1760 and 1780, about six hundred Acts were passed, authorising the construction of new roads and bridges; and by the end of the century, the main roads were in a very fair condition.

The early roads of Scotland and Wales were even worse, if possible, than those of England. Sometimes they lay along the bed of a river. At other times they crossed a barren mountain. In the lower grounds, bogs were always in the way. The Life of Telford will show what that able engineer did towards opening up Scotland and Wales, and making the roads passable, for pleasure or for business, at all times of the year.

The construction of the Highland roads had also an important moral influence. The roads stimulated industry amongst a people who had hitherto been unused to it. In constructing them, the people learnt to work, to use tools, and to apply themselves to continuous labour. Telford himself regarded the Highland roads in the light of a Working Academy, which annually turned out about eight hundred improved workmen.

Down to about the middle of last century, the Highlanders were among the most turbulent of subjects. When not lifting cattle from the Lowlanders, they were engaged in arms against the Government. They levied black-mail all round the Highland frontier as late as 1745. Mr. Grant, of Corrymorry, a Highland chief, used to insist that "a rising in the Highlands was absolutely necessary to give employment to the numerous bands of lawless and idle young men who infested every property." *

During the rebellion of 1745, there was a company of men in one of Prince Charles's regiments, called "The Balmoral Farquharsons." These men fought to the best of their power against her Majesty's forefathers. They were raised from the estate on which the Queen now resides ; and perhaps she has not more peaceable nor affectionate subjects than the Balmoral men, nor than the Highlanders generally, in all her dominions.

After the turnpike roads had been completed—after the hills had been cut down and the valleys raised, so as to level the roads in all directions—after the mails were able to travel at ten miles an hour from London almost to John O'Groat's—and when all complaints as to the quickness of travelling seemed to be removed,—lo! the Locomotive was invented, and all that Telford had done was, to a certain extent, undone by the Stephensons.

One of the last works of Telford was the improvement of the high road between London and Birmingham, so as to enable the coach-road to compete with the rail-road. He was levelling the top of the hill at Weedon, near Daventry, while the railway excavators were working underneath. Near that point, three remarkable things are to be seen, —the remains of Watling Street, the original Roman road,—Telford's last turnpike road,—and the first great rail-road between London and the north.

It will thus be seen, that notwithstanding our

* Anderson's ' Highlands and Islands of Scotland,' p. 434.

delays in opening up our internal communications, we have at last effectually made up our long arrears. We have no longer to send for foreign engineers to execute our necessary works in Embankments and Bridge building. Our native engineers can make docks and harbours without extraneous assistance. We can make our own iron, and our own tools. In fact, instead of borrowing engineers from abroad, we now send them to all parts of the world. British built ships ply on every sea. We export machinery to all countries, and supply Holland itself with pumping engines.

During the last hundred years, our native engineers have completed a magnificent system of canals, turnpike roads, bridges, and railways, by means of which the internal communications of the country have been completely opened up. They have built lighthouses — finger-posts of the sea — round our coasts, by which ships freighted with the produce of all lands are safely lighted to their destined havens. They have hewn out and built docks and harbours for the accommodation of a gigantic commerce ; whilst their inventive genius has rendered iron, fire, and water, the most untiring workers in all branches of industry, and the most effective agents in locomotion by land and sea.

The remarkable thing is, that nearly all this has been accomplished during the last hundred years, and much of it within the life of the present generation.

Smeaton did not erect and light up the first stone lighthouse in Britain, until the year 1759.

Brindley made the first canal for the Duke of Bridgewater in 1761.

Although Watt had been engaged for about ten years in inventing his condensing steam-engine, it was not until 1776 that he erected and sold his first engine. It was patented in 1769, the same year in which Arkwright patented his machine for spinning by rollers.

It was not until the end of last century that England was able to dispense with the supply of the greater part of its iron from abroad; for Henry Cort did not invent his puddling process, which so enormously increased the production of iron, until the years 1783 and 1784.

The first mail coach, on Palmer's system, did not begin to run between London and Bristol until the year 1784.

There were no public Docks in London until the beginning of the present century. The first were constructed by Jessop and Rennie.

The first public exhibition of gas-lighting was made in 1802, when Murdock lit up the front of the Soho manufactory with gas, in celebration of the Peace of Amiens.

Trevithick used his first high-pressure locomotive in 1802, when he ran it as a road locomotive along the streets of his native town.

1802 was an eventful year, for it was then that Symington employed his first successful steamboat on the Forth and Clyde canal.

Fulton's steamboat, with Watt's engine as the motive power, made its first voyage on the Hudson in 1807; and Henry Bell's 'Comet' made its first voyage on the Clyde in 1812; but a considerable time elapsed before steam was generally introduced into either the Commercial or the Royal navy.

In 1819, the 'Savannah' steamboat, of 350 tons,

made the voyage from New York to Liverpool in twenty-six days; and in 1838, the 'Great Western' arrived at New York from Bristol in eighteen days. Since then the Atlantic Ocean has been covered with screw steam-ships; and the direct line between Liverpool and New York has become so crowded, that steam-ships must keep a course so definite that the expression of "steam lanes" has become a matter of ordinary maritime parlance.

George Stephenson invented his Killingworth locomotive for drawing coals in 1814; and his passenger locomotive for the Liverpool and Manchester Railway in 1829.

The first telegraph line was erected on the Great Western Railway, between Paddington and West Drayton, in 1841; and the submarine telegraph was first laid between Dover and Calais in 1851. Since then the telegraph has really "put a girdle round the earth."

The following comprehensive thought occurs in one of Sir Isaac Newton's Reflections:—

" It is clearly apparent that the inhabitants of the world are of a short date, seeing that all arts, as letters, ships, printing, the needle, and such like, have been discovered within the memory of history."

The discoveries of Newton himself were among the greatest of the age he lived in; but since his time various others have been made which strikingly illustrate the truth of his observation.

Only thirty years after his death, Harrison invented the Chronometer, by means of which the navigator can determine his longitude at sea,— an invention almost equal to that of the Mariner's Compass itself.

The science of Chemistry has been discovered,

—dating from the days of Black, Priestley, and Lavoisier; and Natural Philosophy, if not discovered, has at least been greatly developed since Newton's time; while both have been the subject of many inventions of the greatest importance.

Thus the inventions of the steam-engine, the locomotive, and the electric telegraph, have already wonderfully affected the civilised world; and are probably destined to accomplish still more extraordinary changes in the generations that are to come.

One of the most remarkable things about Engineering in England is, that its principal achievements have been accomplished, not by natural philosophers nor by mathematicians, but by men of humble station, for the most part self-educated.

The educated classes of the last century regarded with contempt mechanical subjects. Dean Swift spoke of "that fellow Newton over the way —a glass grinder, and a maker of spectacles." Smeaton was taken to task by his friends of the Royal Society, for having undertaken the "navvy" work of making a road across the valley of the Trent, between Markham and Newark.

At a time when the Court, the Camp, and the Church, formed the principal occupations of the higher classes, engineering was thought unscientific and ungenteel. Hence, when any public work of importance had to be done—some great level of the Fens to be drained, or some harbour to be constructed—a foreign engineer was usually employed to do it.

Nor did any of the great mechanics, who have since invented tools, engines, and machines, at all belong to the educated classes. They received no college education. Some of them could scarcely write their own names. But where learning failed, natural genius triumphed. These men gathered their practical knowledge in the workshop, or acquired it in manual labour. They rose to celebrity, mostly by their habits of observation, their powers of discrimination, their constant self-improvement, and their patient industry.

The greatest stimulus to English engineering has been Trade—the increase of our commerce at home, and the extending of it abroad. England was nothing, compared with continental nations, until she had become commercial. She fought wonderfully, and boasted of many victories; but she was gradually becoming less powerful as a State, until about the middle of last century, when a number of ingenious and inventive men, without apparent relation to each other, arose in various parts of the kingdom, and succeeded in giving an immense impulse to all the branches of the national industry; the result of which has been a harvest of wealth and prosperity, perhaps without a parallel in the history of the world.

Before that time, our commerce had grown very slowly. Spain and Portugal had immense foreign possessions and commerce, when England had none. A small company of English merchants began to trade with India about the beginning of the seventeenth century. The northern colonies of America were founded about the same time; but the government of England afterwards threw them off. Admiral Penn took Jamaica from the Spaniards in 1655; but the other principal islands and possessions in the West Indies were only ceded to England during the wars of the French

Revolution. We obtained possession of the Cape of Good Hope in 1815. New South Wales colony began to be planted in 1788, South Australia in 1836, and Victoria in 1837. The New Zealand Company was only established in 1839.

With the increase of her commerce, about the end of last century, England became more "quick" than she had been before. Intercourse between town and town became more frequent, for the purposes of trade. Then the roads became improved; canals were made; bridges were built; docks and harbours were constructed; and, above all, the steam-engine was invented. Within a few years, Watt's invention created a large number of new industries, and gave employment to immense numbers of people. The engine was perfected while England was engaged in the Revolutionary war with France; and it not only kept our armies in the field, paid the interest upon our debt, but left the nation more powerful than ever it had been before.

At a meeting held in London in 1824, for the purpose of erecting a monument to James Watt,— the greatest of engineers,—Mr. Huskisson used these remarkable words:

"His invention has contributed not only to advance personal comfort and public wealth, by affording to industrious millions the faculty of providing for their industrial wants, by means which directly conduce to the general power and greatness of the state, but also to the general diffusion of a spirit of improvement, a thirst for instruction, and an emulation to apply it to the purposes of practical utility, even in the humblest classes of the community. . . . Looking back, however, to the demands which were made upon the resources of the country during the late war, perhaps it is not too much to say, at least it is my opinion, that these resources might have failed us before that war was brought to a safe and glorious conclusion, but for the creations of Mr. Watt, and of others moving in the same career, by whose discoveries those resources were so greatly multiplied and increased. It is perhaps not too much to say, that but for the vast accession thus imperceptibly made to the general wealth of the empire, we might have been driven to sue for peace, before, in the march and progress of events, Nelson had put forth the last energies of his naval genius at Trafalgar, or, at any rate, before Wellington had put the final seal to the security of Europe at Waterloo."

Watt, however, did far more than assist in the conquests of Trafalgar and Waterloo. He discovered an immense amount of Latent Labour. He yoked his steam-engine to machinery of almost infinite variety, and employed it to create materials of vast importance for the comfort of mankind. The steam-plough and the steam-press are alike its issue. It impels ships, drives locomotives, hammers iron, excavates docks, and, in a word, is the greatest and most indefatigable worker in the world. Even the steam-engine itself was never complete until it could make itself—that is, manufacture the self-acting tools which could never go wrong in the process of constructing it.

One of the circumstances that has particularly attracted the attention of foreigners has been, that almost the whole of our great engineering works have been the spontaneous outgrowths of private and individual industry and genius. Even Watt was a mere tradesman, trying for ten years to work out his great invention; and, but for the noble and generous help of Matthew Boulton, he might never have perfected it.

Government has done next to nothing to promote engineering works. These have been the result of the liberality, public spirit, and commercial enterprise of merchants, traders, and manufacturers. Baron Dupin, in his review of the engineering works of England, observes that "fewer years sufficed for a few individuals to execute and construct, at their private expense, the docks of England, which receive the trading fleets of the two hemispheres, than it required for the triumphant government of France to erect some of the quays of the Seine. These are truly the prodigies of the seas." *

Canova is said to have wondered, while in England, that the trumpery Chinese Bridge, then in St. James's Park, should have been the production of the Government, while Waterloo Bridge was the enterprise of a private company. Baron Dupin, speaking of Rennie's masterpiece, uses these words :—

" If, from the incalculable effect of the Revolutions to which empires are subject, the people of the earth should one day inquire, ' Which was formerly the New Phœnicia and the Western Tyre, which covered the ocean with her vessels ? '—The greater part of her edifices, consumed by a destructive climate, will no longer stand to answer with the dumb language of monuments ; but Waterloo Bridge will ever exist to repeat to the most remote

* 'Mémoires sur la Marine.' 1819.

generations, 'Here stood a rich, industrious, and powerful city!' The traveller at this sight will imagine that some great prince sought to signalise the end of his reign by many years of labour, and to immortalise the glory of his actions by this imposing structure. But if tradition tell him that six years sufficed to begin and complete this work —if he learn that a mere company of merchants built this mass, worthy of Sesostris and the Cæsars—he will the more admire the nation where similar enterprises could be the fruit of the efforts of merchants and private individuals. And if he should then reflect on the causes of the prosperity of empires, he will understand that such a nation must have possessed wise laws, powerful institutions, and a well-protected liberty; for these are stamped in the grandeur and utility of the works completed by her citizens." *

The incidents connected with the Lives of the Engineers themselves are of much interest. In the first place, the early engineers thoroughly represented the national character. Baron Dupin took this view. He said we could not understand the astounding successes of English commerce and the sustained skill and enterprise of English engineers, without taking into consideration that perseverance in all undertakings and that constancy of character on which their successes mainly depended.

There is a promptitude about the Engineer that we do not find in the Architect. The late Prince Consort is reported to have said: " If we want any work done of an unusual character, and send for an architect, he hesitates, debates, trifles : we send for

* Baron Dupin, 'On the Commercial Power of Great Britain,' Vol. i. p. 360. 1825.

an engineer, and he *does it.*" Hence the Prince usually relied on Engineers for carrying out the important works with which he was connected.

The greater number of the early engineers were the offspring of necessity. Some great work required to be done, and forthwith a skilled worker (for there were no so-called engineers in those days) was called upon to do it. The work which he had undertaken to accomplish often presented great difficulties, and his efforts to overcome them amounted to a succession of individual struggles, sometimes rising almost to the heroic. In one case we find the self-born engineer to be a London goldsmith, like Myddelton; in another, a retired sea-captain, like Perry; a wheelwright, like Brindley; an attorney's clerk, like Smeaton; a mathematical instrument maker, like Watt; a millwright, like Rennie; a working mason, like Telford; a slater, like Clement; or an engine brakesman, like Stephenson.

These men were strong-minded, resolute, and ingenious, and were impelled to their special pursuits by the force of their constructive instincts. In most cases they had to make for themselves a way; for there were none to point out the road, which, until they entered upon their undertakings, had for the most part been untravelled. To our mind, there is almost a dramatic interest in their noble efforts, their defeats, their triumphs; and their constant rise, in spite of manifold obstructions and difficulties, from obscurity to fame.

It will scarcely be supposed that men so uneducated as these, had anything to do with the making of the history of England. History is principally monopolised by the deeds and misdeeds of kings, warriors, and statesmen ; and there is little room left for the engineer or the mechanic. But Peace has its battles and its victories as well as War; the results being much more beneficent in one case than they are in the other.

Our engineers may be regarded in some measure as the makers of modern civilisation. The problems of political history cannot properly be interpreted without reference to the people themselves—how they lived and how they worked, and what they did to promote the civilisation of the nation to which they belonged. Hence English engineers are not unworthy to be considered in the history of their country. For what were England without its roads, its bridges, its canals, its docks, and its harbours. What were it without its tools, its machinery, its steam-engine, its steam-ships, and its locomotive. Are not the men who have made the motive power of the country, and immensely increased its productive strength, the men above all others who have tended to make the country what it is ?

The last letter that the author received from Mr. Cobden, contained the following passage :—

" I venture the prediction that not only an enduring but an increasing renown will attach to the memories of those 'Captains of Industry' whose biographies you have recorded; for it cannot be doubted that each succeeding generation will hold in higher estimation those discoveries in physical science to which mankind must attribute henceforth so largely its progress and improvement."

The object of this work is to give an account of some of the principal men by whom this nation has been made so great and prosperous as it is—the men by whose skill and industry large tracts of fertile land have been won from the sea, the bog, and the fen, and made available for human habitation and sustenance ; who, by their industry, skill, and genius, have made England the busiest of workshops ; who have rendered the country accessible in all directions by roads, bridges, canals, and railways ; and who have built lighthouses, breakwaters, docks, and harbours, for the protection and accommodation of our vast home and foreign commerce.

When the author first contemplated writing these memoirs, he was dissuaded rather than encouraged. Robert Stephenson did not think that there was anything in his father's Life to attract the attention of the reading public. " The subject," he said, " possessed no literary interest. What do people care about the building of bridges, the excavation of tunnels, and the making of roads and railways ? And then there was the ' Life of Telford,' which had been published at a great expense, and had fallen still-born from the press."

The author, however, entertained a different opinion. He had lived amongst engineering men for many years. He had himself known George Stephenson, and had heard many interesting things about him from his professional associates ; and he eventually resolved, with the assistance of Robert Stephenson, to proceed with the preparation of his Life. It was published in 1857, and was received by the public with much favour.

The other lives were proceeded with subsequently ; according as time and opportunity permitted. The original work was published at a high price, in consequence of the great expense of the illustrations. But as the edition is now exhausted, the author has been induced to republish the work in a complete form at a reduced price, in the hope that it may thus obtain a more general circulation. In some respects it has been considerably enlarged, and in others compressed ; but no essential feature of the original narrative will be found omitted.

CONTENTS.

------◆------

CHAPTER I.

INTRODUCTORY.

Labour and skill have made modern England—Early settlement of
Britain—Early works of embankment and draining—Reclama-
tion of Romney Marsh—Early embankments of the Thames—
Great Level of the Fens—Laws for preserving the reclaimed
lands—Destructive floods—James I. encourages drainage works
—Cornelius Vermuyden, the Dutch Engineer, employed by King
James—England's dependence upon foreigners in former times
—England's progress in engineering and manufacturing skill

CHAPTER II.

SIR CORNELIUS VERMUYDEN—DRAINAGE OF THE FENS.

Vermuyden's embankment and drainage works at Dagenham and
Windsor—The Isle of Axholme—Reclamation of Hatfield Chase
—The works carried out by foreign capital and foreign labour—
Opposition of the native population—The embankments broken
by the Parliamentarians—Vermuyden undertakes the drainage
of the Great Fen Level—Francis Earl of Bedford becomes chief
undertaker—Opposition of the Fen-men and of the town population—Satirical songs and ballads against the Fen-drainers—
"The Powte's Complaint"—Oliver Cromwell heads an agitation
against the Fen-drainers—The Fen-men destroy the works and
property at Lindsey Level and other Fen districts—Vermuyden's
"discourse" on the drainage of the Fens, 1649—Works completed, 1652—Public thanksgiving—Vermuyden's labours, personal sacrifices, and death—Progressive improvements in the
Fen district—Salubrity of the Fens   ..   ..   ..

CHAPTER III.

SIR HUGH MYDDELTON—CUTTING OF THE NEW RIVER.

CHAPTER IV.

SIR HUGH MYDDELTON (*continued*)—HIS OTHER ENGINEERING AND MINING WORKS—HIS DEATH.

CHAPTER V.

CAPTAIN PERRY—STOPPAGE OF DAGENHAM BREACH.

CHAPTER VI.

THE BEGINNING OF CANAL NAVIGATION.

LIFE OF JAMES BRINDLEY.

CHAPTER I.

BRINDLEY'S EARLY YEARS.

CHAPTER II.

BRINDLEY AS MASTER WHEELWRIGHT AND MILLWRIGHT.

CHAPTER III.

THE DUKE OF BRIDGEWATER—BRINDLEY EMPLOYED AS ENGINEER OF HIS CANAL.

CHAPTER IV.

EXTENSION OF THE DUKE'S CANAL TO THE MERSEY.

CHAPTER V.

THE DUKE'S DIFFICULTIES—COMPLETION OF THE CANAL—GROWTH OF MANCHESTER.

CHAPTER VI.

CHAPTER VII.

APPENDIX.

PIERRE-PAUL RIQUET, CONSTRUCTOR OF THE GRAND CANAL OF LANGUEDOC.

LIST OF ILLUSTRATIONS.

FULL-PAGE PLATES

LINE ILLUSTRATIONS.

JAMES BRINDLEY

AND

THE EARLY ENGINEERS.

CHAPTER I.

INTRODUCTORY.

It has taken the labour and the skill of many generations of men to make England the country that it now is; to reclaim and subdue its lands for purposes of agriculture, to build its towns and supply them with water, to render it easily accessible by means of roads, bridges, canals, and railways, and to construct lighthouses, breakwaters, docks, and harbours for the protection and accommodation of its commerce. Those great works have been the result of the continuous industry of the nation, and the men who have designed and executed them are entitled to be regarded in a great measure as the founders of modern England.

Engineering, like architecture, strikingly marks the several stages which have occurred in the development of society, and throws much curious light upon history. The ancient British encampment, of which many specimens are still to be found on the summits of hills, with occasional

indications of human dwellings within them in
the circular hollows or pits over which huts once
stood,—the feudal castle perched upon its all
but inaccessible rock, provided with drawbridge
and portcullis to secure its occupants against
sudden assault,—then the moated dwelling, situated
in the midst of the champaign country, indicating
a growing, though as yet but half-hearted confidence in the loyalty of neighbours,—and, lastly,
the modern mansion, with its drawing-room windows opening level with the sward of the adjacent country,—all these are not more striking
indications of social progress at the different
stages in our history, than the reclamation and
cultivation of lands won from the sea, the making
of roads and building of bridges, the supplying
of towns with water, and the construction of
canals and railroads for the ready conveyance of
persons and merchandise throughout the empire.

In England, as in all countries, men began
with making provision for food and shelter. The
valleys and low-lying grounds being mostly covered
with dense forests, the naturally cleared high lands,
where timber would not grow, were doubtless
occupied by the first settlers. Tillage was not
as yet understood nor practised; the people sub-
sisted by hunting, or upon their herds of cattle,
which found ample grazing among the hills of
Dartmoor, and on the grassy downs of Wiltshire
and Sussex. Numerous remains or traces of
ancient dwellings have been found in those districts, as at Bowhill in Sussex, along the skirts
of Dartmoor where the hills slope down to the
watercourses, and on the Wiltshire downs, where
Old Sarum, Stonehenge and Avebury, mark the

earliest and most flourishing of the British settlements.

The art of reclaiming, embanking, and draining land, is supposed to have been introduced by men from Belgium and Friesland, who early landed in great numbers along the south-eastern coasts, and made good their footing by the power of numbers, as well as probably by their superior civilisation. The lands from which they came had been won by skill and industry from the sea and from the fen; and when they swarmed over into England, they brought their arts with them. The early settlement of Britain by the races which at present occupy it, is usually spoken of as a series of invasions and conquests; but it is probable that it was for the most part effected by a system of colonisation, such as is going forward at this day in America, Australia, and New Zealand; and that the immigrants from Friesland, Belgium, and Jutland, secured their settlement by the spade far more than by the sword. Wherever the new men came, they settled themselves down on their several bits of land, which became their holdings; and they bent their backs over the stubborn soil, watering it with their sweat; and delved, and drained, and cultivated it, until it became fruitful. They also spread themselves over the richer arable lands of the interior, the older population receding before them to the hunting and pastoral grounds of the north and west. Thus the men of Teutonic race gradually occupied the whole of the reclaimable land, and became dominant, as is shown by the dominancy of their language, until they were stopped by the hills of Cumberland, of Wales, and of Cornwall. The same process seems to have gone on in the arable districts of Scotland, into which a swarm of colonists from Northumberland poured in the reign of David I., and quietly settled upon the soil, which they proceeded to cultivate. It is a remarkable confirmation of this view of the early settlement of the country by its present races, that the modern English language extends over the whole of the arable land of England and Scotland, and the Celtic tongue only begins where the plough ends.

One of the most extensive districts along the English coast, lying the nearest to the country from which the continental immigrants first landed, was the tract of Romney Marsh, containing about 60,000 acres of land along the south coast of Kent. The reclamation of this tract is supposed to be due to the Frisians. English history does not reach so far back as the period at which Romney Marsh was first reclaimed, but doubtless the work is one of great antiquity. The district is about fourteen miles long and eight broad, divided into Romney Marsh, Wallend Marsh, Denge Marsh, and Guildford Marsh. The tract is a dead, uniform level, extending from Hythe, in Kent, westward to Winchelsea, in Sussex; and it is to this day held from the sea by a continuous wall or bank, on the solidity of which the preservation of the district depends, the surface of the marsh being under the level of the sea at the highest tides. The following descriptive view of the marsh, taken from the high ground above the ancient Roman fortress of Portus Limanis, near the more modern but still ancient castle of Lymne, will give an idea of the extent and geographical relations of the district.

The tract is so isolated, that the marshmen say the world is divided into Europe, Asia, Africa, America, and Romney Marsh. It contains few or no trees, its principal divisions being formed by dykes and watercourses. It is thinly peopled, but abounds in cattle and sheep of a peculiarly hardy breed, which are a source of considerable wealth

Descriptive View of Romney Marsh, from Lymne Castle.

to the marshmen; and it affords sufficient grazing for more than half a million of sheep, besides numerous herds of cattle.

The first portion of the district reclaimed was an island, on which the town of Old Romney now stands; and embankments were extended southward as far as New Romney, where an accumulation of beach took place, forming a natural barrier

against further encroachments of the sea at that
point. The old town of Lydd originally stood
upon another island, as did Ivychurch, Old Winchelsea, and Guildford; the sea sweeping round
them and rising far inland at every tide. Bur-
marsh, and the districts thereabout, were reclaimed
at a more recent period; and by degrees the
islands disappeared, the sea was shut out, and
the whole became firm land. Large additions
were made to it from time to time by the deposits
of shingle along the coast, which left several
towns, formerly important seaports, stranded upon
the beach far inland. Thus the ancient Roman
port at Lymne, past which the Limen or Rother
is supposed originally to have flowed, is left high
and dry more than three miles from the sea, and
sheep now graze where formerly the galleys of
the Romans rode. West Hythe, one of the Cinque
Ports, originally the port for Boulogne, is silted
up by the wide extent of shingle used by the
modern School of Musketry as their practising-
ground. Old Romney, past which the Rother
afterwards flowed, was one of the ancient ports
of the district, but it is now about two miles
from the sea. The marshmen followed up the
receding waters, and founded the town of New
Romney, which also became a Cinque Port; but
a storm that occurred in the reign of Edward I.
so blocked up the Rother with shingle, at the
same time breaching the wall, that the river took
a new course, and flowed thenceforward by Rye
into the sea; and the port of New Romney became
lost. The point of Dungeness, running almost
due south, gains accumulations of shingle so
rapidly from the sea, that it is said to have

extended more than a mile seaward within the
memory of persons living. Rye was founded on
the ruins of the Romneys, and also became a
Cinque Port ; but notwithstanding the advantage
of the river Rother flowing past it, that port also
has become nearly silted up, and now stands
about two miles from the sea. New Winchelsea,
the Portsmouth and Spithead of its day, is left
stranded like the rest of the Old Cinque Ports,
and is now but a village surrounded by the remains of its ancient grandeur. All this ruin,
however, wrought by the invasions of the shingle
upon the seacoast towns, has only served to increase the area of the rich grazing ground of the
marsh, which continues year by year to extend
itself seaward.

Another highly important work of the same
class was the embankment of the Thames, now
the watery highway between the capital of Great
Britain and the world. Before human industry had
confined the river within its present channel, it
was a broad estuary, in many parts between London
and Gravesend several miles wide. The higher
tides covered Plumstead and Erith Marshes on the
south, and Plaistow, East Ham, and Barking Levels
on the north ; the river meandering in many devious
channels at low water, leaving on either side vast
expanses of rich mud and ooze. Opposite the City
of London, the tides washed over the ground now
covered by Southwark and Lambeth ; the district
called Marsh still reminding us of its former state,
as Bankside informs us of the mode by which it
was reclaimed by the banking out of the tidal waters.

A British settlement is supposed to have been
formed at an early period on the high ground on

which St. Paul's Cathedral stands, by reason of its
natural defences, being bounded on the south by
the Thames, on the west by the Fleet, and on the
north and east by morasses, Moorfields Marsh

Valley of the Thames (Western Part). [From the Ordnance Survey.]

having only been reclaimed within a comparatively
recent period. The natural advantages of the situation were great, and the City seems to have acquired
considerable importance even before the Roman
period. The embanking of the river has been attributed to that indefatigable people; but on this
point no evidence exists. The numerous ancient
British camps found in all parts of the kingdom
afford sufficient proof that the early inhabitants of
the country possessed a knowledge of the art of
earthwork; and it is not improbable that the same
Belgian tribes who reclaimed Romney Marsh were
equally quick to detect the value for agricultural
purposes of the rich alluvial lands along the valley
of the Thames, and proceeded accordingly to embank them after the practice of the country from
which they had come. The work was carried on
from one generation to another, as necessity required, until the Thames was confined within its
present limits, the process of embanking serving to

deepen the river and improve it for purposes of navigation, while large tracts of fertile land were at the same time added to the food-producing capacity of the country.

Map of the Valley of the Thames (Eastern Part).
N.B.—The dotted line represents the embankments raised along the banks of the river.

Another of the districts won from the sea, in which a struggle of skill and industry against the power of water, both fresh and salt, has been persistently maintained for centuries, is the extensive low-lying tract of country, situated at the junction of the counties of Lincoln, Huntingdon, Cambridge, and Norfolk, commonly known as the Great Level of the Fens. The area of this district presents almost the dimensions of a province, being from sixty to seventy miles from north to south, and from twenty to thirty miles broad, the high lands of the interior bounding it somewhat in the form of a horse-shoe. It contains about 680,000 acres of the richest land in England, and is as much the product of art as the kingdom of Holland, opposite to which it lies.

Not many centuries ago, this vast tract of about two thousand square miles of land was entirely abandoned to the waters, forming an immense estuary of the Wash, into which the rivers Witham,

Welland, Glen, Nene, and Ouse discharged the
rainfall of the central counties of England. It was
an inland sea in winter, and a noxious swamp in
summer, the waters expanding in many places
into settled seas or meres, swarming with fish and
screaming with wild-fowl. The more elevated parts

Map of the Fens as they lay drowned. (After Dugdale.)

were overgrown with tall reeds, which appeared at
a distance like fields of waving corn; and they were
haunted by immense flocks of starlings, which,
when disturbed, would rise in such numbers as
almost to darken the air. Into this great dismal
swamp the floods descending from the interior

were carried, their waters mingling and winding
by many devious channels before they reached
the sea. They were laden with silt, which became
deposited in the basin of the Fens. Thus the river-
beds were from time to time choked up, and the
intercepted waters forced new channels through
the ooze, meandering across the level, and often
winding back upon themselves, until at length the
surplus waters, through many openings, drained
away into the Wash. Hence the numerous abandoned beds of old rivers still traceable amidst the
Great Level of the Fens—the old Nene, the old
Ouse, and the old Welland. The Ouse, which in
past times flowed into the Wash at Wisbeach (or
Ouse Beach), now enters at King's Lynn, near
which there is another old Ouse. But the probability is that all the rivers flowed into a lake,
which existed on the tract known as the Great
Bedford Level, from thence finding their way, by
numerous and frequently shifting channels, into
the sea.

Along the shores of the Wash, where the fresh
and salt waters met, the tendency to the deposit of
silt was the greatest ; and in the course of ages, the
land at the outlets of the inland waters became
raised above the level of the interior. Accordingly,
the first land reclaimed in the district was the rich
fringe of deposited silt lying along the shores of
the Wash, now known as Marshland and South
Holland. This was effected by the Romans, a hard-
working, energetic, and skilful people ; of whom
the Britons are said to have complained * that they
wore out and consumed their hands and bodies
in clearing the woods and banking the fens. The

* Tacitus, 'Life of Agricola.'

bulwarks or causeways which they raised to keep
out the sea are still traceable at Po-Dyke in Marshland, and at various points near the old coast-line.
On the inland side of the Fens the Romans are supposed to have constructed another great work of
drainage, still known as Carr Dyke, extending from
the Nene to the Witham. It means Fen Dyke, the
fens being still called Carrs in certain parts of
Lincoln. This old drain is about sixty feet wide,
with a broad, flat bank on each side; and originally it must have been at least forty miles in
extent, winding along under the eastern side of
the high land, which extends in an irregular line
up the centre of the district from Stamford to
Lincoln.

The eastern parts of Marshland and Holland
were thus the first lands reclaimed in the Level,
and they were available for purposes of agriculture
long before any attempts had been made to drain
the lands of the interior. Indeed, it is not improbable that the early embankments thrown up along
the coast had the effect of increasing the inundations of the lower-lying lands farther west; for,
whilst they dammed the salt water out, they also
held back the fresh, no provision having been made
for improving and deepening the outfalls of the
rivers flowing through the Level into the Wash.
The Fen lands in winter were thus not only
flooded by the rainfall of the Fens themselves,
and by the upland waters which flowed from the
interior, but also by the daily flux of the tides
which drove in from the German Ocean, holding
back the fresh waters, and even mixing with them
far inland.

The Fens, therefore, continued flooded with

water down to the period of the Middle Ages, when there was water enough in the Witham to float the ships of the Danish sea rovers as far inland as Lincoln, where ships' ribs and timbers have recently been found deep sunk in the bed of the river. The first reclaimers of the Fen lands seem to have been the religious recluses, who settled upon the islands overgrown with reeds and rushes, which rose up at intervals in the Fen level, and where they formed their solitary settlements. One of the first of the Fen islands thus occupied was the Isle of Ely, or Eely—so called, it is said, because of the abundance and goodness of the eels caught in the neighbourhood, and in which rents were paid in early times. It stood solitary amid the waste of waters, and was literally an island. Etheldreda, afterwards known as St. Audrey, the daughter of the King of the East Angles, retired thither, secluding herself from the world and devoting herself to a recluse life. A nunnery was built, then a town, and the place became famous in the religious world. The pagan Danes, however, had no regard for Christian shrines, and a fleet of their pirate ships, sailing across the Fens, attacked the island and burnt the nunnery. It was again rebuilt, and a church sprang up, the fame of which so spread abroad that Canute, the Danish king, determined to visit it. It is related that as his ships sailed towards the island his soul rejoiced greatly, and on hearing the chanting of the monks in the quire wafted across the waters, the king joined in the singing and ceased not until he had come to land. Canute more than once sailed across the Fens with his ships, and the tradition survives that on one occasion, when passing from Ramsey to Peterborough, the waves were so boisterous on Whittlesea Mere (now a district of fruitful cornfields), that he ordered a channel to be cut through the body of the Fen westward of Whittlesea to Peterborough, which to this day is called by the name of the " King's Delph."

The other Fen islands which became the centres of subsequent reclamations were Crowland, Ramsey, Thorney, and Spinney, each the seat of a monastic establishment. The old churchmen, notwithstanding their industry, were, however, only able to bring into cultivation a few detached points, and made very little impression upon the drowned lands of the Great Level. It often happened, indeed, that the steps which they took to drain one spot merely had the effect of sending an increased flood of water upon another, and perhaps diverting in some new direction the water which before had driven a mill, or formed a channel for purposes of navigation. The rivers also were constantly liable to get silted up, and form for themselves new courses ; and sometimes, during a high tide, the sea would burst in, and in a single night undo the tedious industry of centuries.

Each suffering locality, acting for itself, did what it could to preserve the land which had been won, and to prevent the recurrence of inundations. Dyke-reeves were appointed along the sea-borders, with a force of shore-labourers at their disposal, to see to the security of the embankments ; and fen-wards were constituted inland, over which commissioners were set, for the purpose of keeping open the drains, maintaining the dykes, and preventing destruction of life and property by floods, whether descending into the Fens from the high lands or bursting in upon them from the sea. Where lands became suddenly drowned, the Sheriff was authorised to impress diggers and labourers for raising embankments; and commissioners of sewers were afterwards appointed, with full powers of local action, after the law and usage of Romney Marsh. In one district we find a public order made that every man should plant with willows the bank opposite his portion of land towards the fen, "so as to break off the force of the waves in flood times;" and swine were not to be allowed to go upon the banks unless they were ringed, under a penalty of a penny (equal to a shilling in our money) for every hog found unringed. A still more terrible penalty for neglect is mentioned by Harrison, who says, "Such as having walls or banks near unto the sea, and do suffer the same to decay (after convenient admonition), whereby the water entereth and drowneth up the country, are by a certain ancient custom apprehended, condemned, and *staked in the breach*, where they remain for ever as parcel of the new wall that is to be made upon them, as I have heard reported." *

The Great Level of the Fens remained in a comparatively unreclaimed state down even to the end of the sixteenth century; and constant inundations took place, destroying the value of the little settlements which had by that time been won from the watery waste. It would be difficult to imagine anything more dismal than the aspect which the Great Level then presented. In winter, a sea without waves; in summer, a dreary mud-swamp. The atmosphere was heavy with pestilential vapours, and swarmed with insects. The meres and pools were, however, rich in fish and

* Harrison's Preface to Holinshed's Chronicle, i. 313.

wild-fowl. The Welland was noted for stickle-
backs, a little fish about two inches long, which
appeared in dense shoals near Spalding every
seventh or eighth year, and used to be sold
during the season at a halfpenny a bushel, for
field manure. Pike was plentiful near Lincoln :
hence the proverb, "Witham pike, England hath
none like." Fen-nightingales, or frogs, especially
abounded. The birds-proper were of all kinds ;
wild-geese, herons, teal, widgeons, mallards, grebes,
coots, godwits, whimbrels, knots, dottrels, yelpers,
ruffs, and reeves, some of which have long since
disappeared from England. Mallards were so
plentiful that 3000 of them, with other birds in
addition, have been known to be taken at one
draught. Round the borders of the Fens there
lived a thin and haggard population of "Fen-
slodgers," called "yellow-bellies" in the inland
counties, who derived a precarious subsistence
from fowling and fishing. They were described
by writers of the time as "a rude and almost
barbarous sort of lazy and beggarly people."
Disease always hung over the district, ready to
pounce upon the half-starved fenmen. Camden
spoke of the country between Lincoln and Cambridge as "a vast morass, inhabited by fenmen,
a kind of people, according to the nature of
the place where they dwell, who, walking high
upon stilts, apply their minds to grazing, fishing,
or fowling." The proverb of "Cambridgeshire
camels" doubtless originated in this old practice
of stilt-walking in the Fens ; the fenmen, like the
inhabitants of the Landes, mounting upon high
stilts to spy out their flocks across the dead level.
But the flocks of the fenmen consisted principally of

geese, which were called the "fenmen's treasure;" the fenman's dowry being "three-score geese and a pelt" or sheep-skin used as an outer garment. The geese throve where nothing else could exist, being equally proof against rheumatism and ague, though lodging with the natives in their sleeping-places. Even of this poor property, however, the slodgers were liable at any time to be stripped by sudden inundations.

In the oldest reclaimed district of Holland, containing many old village churches, the inhabitants, in wet seasons, were under the necessity of rowing to church in their boats. In the other less reclaimed parts of the Fens the inhabitants were much worse off. "In the winter time," said Dugdale, "when the ice is only strong enough to hinder the passage of boats, and yet not able to bear a man, the inhabitants upon the hards and banks within the Fens can have no help of food, nor comfort for body or soul; no woman aid in her travail, no means to baptize a child or partake of the Communion, nor supply of any necessity saving what these poor desolate places do afford. And what expectation of health can there be to the bodies of men, where there is no element good? the air being for the most part cloudy, gross, and full of rotten harrs; the water putrid and muddy, yea, full of loathsome vermin; the earth spungy and boggy, and the fire noisome by the stink of smoaky hassocks."

The wet character of the soil at Ely may be inferred from the circumstance that the chief crop grown in the neighbourhood was willows; and it was a common saying there, that "the profit of willows will buy the owner a horse before that

by any other crop he can pay for his saddle."
There was so much water constantly lying above
Ely, that in olden times the Bishop of Ely was
accustomed to go in his boat to Cambridge. When
the outfalls of the Ouse became choked up by
neglect, the surrounding districts were subject to
severe inundations; and after a heavy fall of rain,
or after a thaw in winter, when the river swelled
suddenly, the alarm spread abroad, "the bailiff
of Bedford is coming!" the Ouse passing by that
town. But there was even a more terrible visitor
than the bailiff of Bedford; for when a man was
stricken down by the ague, it was said of him,
"he is arrested by the bailiff of Marsh-land;" this
disease extensively prevailing all over the district
when the poisoned air of the marshes began to
work.

The great perils which constantly threatened
the district at length compelled the attention of
the legislature. In 1607, shortly after the accession
of James I., a series of destructive floods burst in
the embankments along the east coast, and swept
over farms, homesteads, and villages, drowning
large numbers of people and cattle. When the
King was informed of the great calamity which
had befallen the inhabitants of the Fens, principally through the decay of the old works of
drainage and embankment, he is said to have
made the right royal declaration, that "for the
honour of his kingdom, he would not any longer
suffer these countries to be abandoned to the will
of the waters, nor to let them lie waste and unprofitable; and that if no one else would undertake their drainage, he himself would become
their undertaker." A Commission was appointed

to inquire into the extent of the evil, from which
it appeared that there were not less than 317,242
acres of land lying outside the then dykes which
required drainage and protection. A bill was
brought into Parliament to enable rates to be
levied for the drainage of this land, but it was
summarily rejected. Two years later, a "little
bill," for draining 6000 acres in Waldersea County,
was passed—the first district Act for Fen drainage
that received the sanction of Parliament. The
King then called Chief-Justice Popham to his aid,
and sent him down to the Fens to undertake a
portion of the work ; and he induced a company
of Londoners to undertake another portion, the
adventurers receiving two-thirds of the reclaimed
lands as a recompense. "Popham's Eau," and
"The Londoners' Lode," still mark the scene of
their operations. The works, however, did not
prove very successful, not having been carried
out with sufficient practical knowledge on the part
of the adventurers, nor after any well-devised plan.
There were loud calls for some skilled undertaker
or engineer (though the latter word was not then
in use) to stay the mischief, reclaim the drowned
lands, and save the industrious settlers in the
Fens from total ruin. But no English engineer
was to be found ready to enter upon so large an
undertaking ; and in his dilemma the King called
to his aid one Cornelius Vermuyden, a Dutch
engineer, a man well skilled in works of embanking
and draining.

The necessity for employing a foreign engineer
to undertake so great a national work is sufficiently
explained by the circumstance that England was
then very backward in all enterprises of this sort.

We had not yet begun that career of industrial skill in which we have since achieved so many triumphs, but were content to rely mainly upon the assistance of foreigners. Holland and Flanders supplied us with our best mechanics and engineers. Not only did Vermuyden prepare the plans and superintend the execution of the Great Level drainage, but the works were principally executed by Flemish workmen. Many other foreign "adventurers" as they were called, besides Vermuyden, carried out extensive works of reclamation and embankment of waste lands in England. Thus a Fleming named Freeston reclaimed the extensive marsh near Wells in Norfolk; Joas Croppenburgh and his company of Dutch workmen reclaimed and embanked Canvey Island near the mouth of the Thames; Cornelius Vanderwelt, another Dutchman, enclosed Wapping Marsh by means of a high bank, along which a road was made, called "High Street" to this day; while two Italians, named Acontius and Castilione, reclaimed the Combe and East Greenwich marshes on the south bank of the river.

We also relied very much on foreigners for our harbour engineering. Thus, when a new haven was required at Yarmouth, Joas Johnson, the Dutchman, was employed to plan and construct it. When a serious breach occurred in the banks of the Witham at Boston, Mathew Hakes was sent for from Gravelines, in Flanders, to repair it; and he brought with him not only the mechanics, but the manufactured iron required for the work. In like manner, any unusual kind of machinery was imported from Holland or Flanders ready made. When an engine was needed to pump water from the Thames for the supply of London, Peter Morice, the Dutchman, brought one from Holland, together with the necessary workmen.

England was in former times regarded principally as a magazine for the supply of raw materials, which were carried away in foreign ships, and returned to us worked up by foreign artisans. We grew wool for Flanders, as India, America, and Egypt grow cotton for England now. Even the wool manufactured at home was sent to the Low Countries to be dyed. Our fisheries were so unproductive, that the English markets were supplied by the Dutch, who sold us the herrings caught in our own seas, off our own shores. Our best ships were built for us by Danes and Genoese ; and when any skilled sailors' work was wanted, foreigners were employed. Thus, when the 'Mary Rose' sank at Spithead in 1545, Peter de Andreas, the Venetian, with his ship carpenter and three Italian sailors, were employed to raise her, sixty English mariners being appointed to attend upon them merely as labourers.

In short, we depended for our engineering, even more than we did for our pictures and our music, upon foreigners. Nearly all the continental nations had a long start of us in art, in science, in mechanics, in navigation, and in engineering. At a time when Holland had completed its magnificent system of water communication, and when France, Germany, and even Russia had opened up important lines of inland navigation, England had not cut a single canal, whilst our roads were about the worst in Europe.

CHAPTER II.

CORNELIUS VERMUYDEN, the Dutch engineer, was invited over to England about the year 1621, to stem a breach in the Thames embankment near Dagenham, which had been burst through by the tide. He was a person of good birth and education, and was born at St. Martin's Dyke, in the island of Tholen, in Zealand. He had been trained as an engineer, and having been brought up in a district where embanking was studied as a profession, and gave employment to a large number of persons, he was familiar with the most approved methods of protecting land against the encroachments of the sea. He was so successful in his operations at Dagenham, that when it was found necessary to drain the Royal park at Windsor, he was employed to conduct the work; and he thus became known to the King, who shortly after employed him in the drainage of Hatfield Level, then a royal chase on the borders of Yorkshire.

The extensive district of Axholme, of which Hatfield Chase formed only a part, resembled the Great Level of the Fens in many respects, being a large fresh-water bay formed by the confluence of the rivers Don, Went, Ouse, and Trent, which brought down into the Humber almost the entire rainfall of Yorkshire, Derbyshire, Nottingham, and North Lincoln, and into which the sea also washed. The uplands of Yorkshire bounded this watery tract on the west, and those of Lincolnshire on the east. Rising up about midway between them was a single hill, or rather elevated ground, formerly an island, and still known as the Isle of Axholme. There was a ferry between Sandtoft and that island in times not very remote, and the farmers of Axholme were accustomed to attend market at Doncaster in their boats, though the bottom of the sea over which they then rowed is now amongst the most productive corn-land in England. The waters extended to Hatfield, which lies along the Yorkshire edge of the level on the west; and it is recorded in the ecclesiastical history of that place that a company of mourners, with the corpse they carried, were once lost when proceeding by boat from Thorne to Hatfield. When Leland visited the county in 1607, he went by boat from Thorne to Tudworth, over what at this day is rich ploughed land. The district was marked by numerous merestones, and many fisheries are still traceable in local history as having existed at places now far inland.

The Isle of Axholme was in former times a stronghold of the Mowbrays, being unapproachable save by water. In the reign of Henry II., when Lord Mowbray held it against the King, it was taken by the Lincolnshire men, who attacked it in boats; and, down to the reign of James I., the only green spot which rose above the wide waste of waters was this solitary isle. Before that monarch's time the south-eastern part of the county of York, from Conisborough Castle to the sea, belonged for the most part to the crown; but

one estate after another was alienated, until at
length, when James succeeded to the throne of
England, there only remained the manor of Hatfield, which, watery though it was, continued to
be dignified with the appellation of a Royal Chase.
There was, however, plenty of deer in the neighbourhood, for De la Pryme says that in his time
they were as numerous as sheep on a hill, and that
venison was as abundant as mutton in a poor man's
kitchen.*   But the principal sport which Hatfield
furnished was in the waters and meres adjacent to
the old timber manor-house.   Prince Henry, the
King's eldest son, on the occasion of a journey to
York, rested at Hatfield on his way, and had a day's
sport in the Royal Chase, which is thus described
by De la Pryme :—" The prince and his retinue all
embarked themselves in almost a hundred boats that
were provided there ready, and having frightened
some five hundred deer out of the woods, grounds,
and closes adjoining, which had been drawn there
the night before, they all, as they were commonly
wont, took to the water, and this little royal navy
pursuing them, soon drove them into that lower
part of the level, called Thorne Mere, and there,
being up to their very necks in water, their horned
heads raised themselves so as almost to represent
a little wood.   Here being encompassed about with
the little fleet, some ventured amongst them, and
feeling such and such as were fattest, they either
immediately cut their throats, or else tying a strong
rope to their heads, drew them to land and killed
them."

Such was the last battue in the Royal Chase of
Hatfield.   Shortly after, King James brought the

* De la Pryme, 'History of the Level of Hatfield Chase.'

subject of the drainage of the tract under the notice
of Cornelius Vermuyden, who, on inspecting it,
declared the project to be quite practicable. The
level of the Chase contained about 70,000 acres, the
waters of which, like those of the Fens, found their
way to the sea through many changing channels.
Various attempts had been made to diminish the
flooding of the lands. In the fourteenth century
several deep trenches were dug, to let off the
water, but they probably admitted as much as
they allowed to escape, and the drowning continued. Commissioners were appointed, but they
did nothing. The country was too poor, and the
people too unskilled, to undertake so expensive and
laborious an enterprise as the effectual drainage
of so large a tract.

A local jury was summoned by the King to
consider the question, but they broke up, after
expressing their opinion of the utter impracticability of carrying out any effective plan for the
withdrawal of the waters. Vermuyden, however,
declared that he would undertake and bind himself
to do that which the jury had pronounced to be
impossible. The Dutch had certainly been successful beyond all other nations in projects of the
same kind. No people had fought against water
so boldly, so perseveringly, and so successfully.
They had made their own land out of the mud of
the rest of Europe, and, being rich and prosperous,
were ready to enter upon similar enterprises in
other countries. On the death of James I., his
successor confirmed the preliminary arrangement
which had been made with Vermuyden, with a
view to the drainage of Hatfield Manor; and on the
24th of May, 1626, after a good deal of negotiation

as to terms, articles were drawn up and signed between the Crown and Vermuyden, by which the latter undertook to reclaim the drowned lands, and make them fit for tillage and pasturage. It was a condition of the contract that Vermuyden and his partners in the adventure were to have granted to them one entire third of the lands so recovered from the waters.

Vermuyden was a bold and enterprising man, full of energy and resources. He also seems to have possessed the confidence of capitalists in his own country, for we find him shortly after proceeding to Amsterdam to raise the requisite money, of which England was then so deficient; and a company was formed composed almost entirely of Dutchmen, for the purpose of carrying out the necessary works of reclamation. Amongst those early speculators in English drainage we find the names of the Valkenburgh family, the Van Peenens, the Vernatti, Andrew Boccard, and John Corsellis. Of the whole number of shareholders amongst whom the lands were ultimately divided, the only names of English sound are those of Sir James Cambell, Knight, and Sir John Ogle, Knight, who were about the smallest of the participants.

Several of the Dutch capitalists came over in person to look after their respective interests in the concern, and Vermuyden proceeded to bring together from all quarters a large number of workmen, mostly Dutch and Flemish. It so happened that there were then settled in England numerous foreign labourers—Dutchmen who had been brought from Holland to embank the lands at Dagenham and Canvey Island on the Thames, and others who had been driven from their own countries by religious persecution—French Protestants from Picardy, and Walloons from Flanders. The countries in which those people had been born and bred resembled in many respects the marsh and fen districts of England, and they were practically familiar with the reclamation of such lands, the digging of drains, the raising of embankments, and the cultivation of marshy ground. Those immigrants had already settled down in large numbers in the eastern counties, and along the borders of the Fens, at Wisbeach, Whittlesea, Thorney, Spalding, and the neighbourhood.* The poor foreigners readily answered Vermuyden's call, and many of them took service under him at Hatfield Chase, where they set to work with such zeal, and laboured with such diligence, that before the end of the second year the work was so far advanced, that a commission was issued for the survey and division amongst the participants of the reclaimed lands.

The plan of drainage adopted seems to have been, to carry the waters of the Idle by direct channels into the Trent, instead of allowing them to meander at will through the level of the Chase. Deep drains were cut, through which the water was drawn from the large pools standing near Hatfield and Thorne. The Don also was blocked out of the level by embankments, and forced

---

* It has been observed that the buildings in many of the old Fen towns to this day have a Flemish appearance, as the names of many of the inhabitants have evidently a foreign origin. Those of Descow, Le Plas, Egar, Bruynne, &c., are said to be still common. Among the settlers in the level of Hatfield was Mathew de la Pryme, who emigrated from Ypres in Flanders during the persecutions of the Duke of Alva. The Prymes of Cambridge are lineally descended from him. Tablets to several of the name are still to be found in Hatfield Church.

through its northern branch, by Turn-bridge, into the river Aire. But this last attempt proved a mistake, for the northern channel was found insufficient for the discharge of the waters, and

Map of the Level of Hatfield Chase. [Corrected, after Dugdale.]

floodings of the old lands about Fishlake, Sykehouse, and Snaith took place; to prevent which, a wide and deep channel, called the Dutch River, was afterwards cut, and the waters of the Don were sent directly into the Ouse, near Goole. This great and unexpected addition to the cost of the undertaking appears to have had a calamitous effect, and brought distress and ruin on many who had engaged in it. The people who dwelt on the northern branch of the Don complained loudly of the adventurers, who were denounced as foreigners and marauders ; and they were not satisfied with mere outcry, but took the law into their own hands ; broke down the embankments, assaulted the Flemish workmen, and several persons lost their lives in the course of the riots which ensued.*

Vermuyden did what he could to satisfy the inhabitants. He employed large numbers of native workmen, at considerably higher wages than had before been paid ; and he strenuously exerted himself to relieve those who had suffered from the changes he had effected, so far as could be done without incurring a ruinous expense.† Dugdale

---

* R. Ansbie writes the Duke of Buckingham from Tickill Castle, under date the 21st August, 1628, as follows :—"What has happened betwixt Mr. Vermuyden's friends and workmen and the people of the Isle of Axholme these inclosed will give a taste. Great riots have been committed by the people, and a man killed by the Dutch party, the killing of whom is conceived to be murder in all who gave direction for them to go armed that day. These outrages will produce good effects. They will procure conformity in the people, and enforce Vermuyden to sue for favour at the Duke's hands,—if not for himself, for divers of his friends, especially for Mr. Saines, a Dutchman, who has an adventure of 13,000l. in this work. Upon examination of the rest of Vermuyden's people, thinks it will appear that he gave them orders to go armed."— 'State Papers,' vol. cxiii. 38.

† F. Vernatti, one of the Dutch capitalists who had contributed largely towards the cost of the works, writes to Monsieur St. Gillis, in October, 1628 :—"The absence of Mr. Vermuyden, and the great interest the writer takes in the business of embankment at

relates that there could be no question about the great benefits which the execution of the drainage works conferred upon the labouring population; for whereas, before the reclamation, the country round about had been "full of wandering beggars," these had now entirely disappeared, and there was abundant employment for all who would work, at good wages. An immense tract of rich land had been completely recovered from the waters, but it could only be made valuable and productive after long and diligent cultivation.

Vermuyden was throughout well supported by the Crown, and on the 6th of January, 1629, he received the honour of knighthood at the hands of Charles I., in recognition of the skill and energy which he had displayed in adding so large a tract to the cultivable lands of England. In the same year he took a grant from the Crown of the whole of the reclaimed lands in the manor of Hatfield, amounting to about 24,500 acres, agreeing to pay the Crown the sum of 16,080*l.*, an annual rent of 193*l.* 3*s.* 5½*d.*, one red rose ancient rent, and an improved rent of 425*l.* from Christmas, 1630.* Power was also granted him to erect one or more chapels wherein the Dutch and Flemish settlers might worship in their own language. They built houses, farmsteads, and windmills; intending to

Haxey, has led him to engage in it with eye and hand. The mutinous people have not only desisted from their threats, but now give their work to complete the dyke, which they have fifty times destroyed and thrown into the river. A royal proclamation made by a serjeant-at-arms in their village, accompanied by the sheriff and other officials, with fifty horsemen, and an exhortation mingled with threats of fire and vengeance, have produced this result."— 'State Papers,' vol. cxix. 73.

* 'State Papers,' vol. cxlvii. 21.

settle down peacefully to cultivate the soil which
their labours had won.

It was long, however, before the hostility and
jealousy of the native population could be ap-
peased. The idea of foreigners settling as colonists
upon lands over which, though mere waste and
swamp, their forefathers had enjoyed rights of
common, was especially distasteful to them, and
bred bitterness in many hearts. The dispossessed
fenmen had numerous sympathisers among the rest
of the population. Thus, on one occasion, we find
the Privy Council sending down a warrant to all
Postmasters to furnish Sir Cornelius Vermuyden
with horses and a guide to enable him to ride post
from London to Boston, and from thence to Hatfield.* But at Royston "Edward Whitehead, the
constable, in the absence of the postmaster, refused
to provide horses, and on being told he should
answer for his neglect, replied, 'Tush! do your
worst : you shall have none of my horses in spite
of your teeth.'"† Complaints were made to the
Council of the injury done to the surrounding
districts by the drainage works ; and an inquisition
was held on the subject before the Earls of Clare
and Newcastle, and Sir Gervase Clifton. Vermuyden was heard in defence, and a decision was given
in his favour ; but he seems to have acted with pre-
cipitancy in taking out subpœnas against many of
the old inhabitants for damage said to have been
done to him and his agents. Several persons were
apprehended and confined in York gaol, and the

* Warrant of Council, dated Whitehall, May 12th, 1630.—
'State Papers,' vol. clxvi. 56.

† Affidavit of George Johnson, servant of Sir Cornelius Vermuyden —'State Papers,' vol. clxx. 17.

feeling of bitterness between the native population and the Dutch settlers grew more intense from day to day. Lord Wentworth, President of the North, at length interfered ; and after surveying the lands, he ordered that all suits should cease. Vermuyden was also directed to assign to the tenants certain tracts of moor and marsh ground, to be enjoyed by them in common. He attempted to evade the decision, holding it to be unjust ; but the Lord President was too powerful for him, and feeling that further opposition was of little use, he resolved to withdraw from the undertaking, which he did accordingly ; first conveying his lands to trustees, and afterwards disposing of his interest in them altogether.*

The necessary steps were then taken to relieve the old lands which had been flooded, by the cutting of the Dutch River at a heavy expense. Great difficulty was experienced in raising the requisite funds ; the Dutch capitalists now holding their hand, or transferring their interest to other proprietors, at a serious depreciation in the value of their shares. The Dutch River was, however, at

* The Dutch settlers lived for the most part in single houses, dispersed through the newly-recovered country. A house built by Vermuyden remains. It was chiefly of timber, and what is called *stud-bound*. It was built round a quadrangular court. The eastern front was the dwelling-house. The other three sides were stables and barns. Another good house was built by Mathew Valkenburgh, on the Middle Ing, near the Don, which afterwards became the property of the Boynton family. Sir Philibert Vernatti and the two De Witts erected theirs near the Idle. A chapel for the settlers was also erected at Sandtoft, in which the various ordinances of religion were performed, and the public service was read alternately in the Dutch and French languages.—The Rev. Joseph Hunter's 'History and Topography of the Deanery of Doncaster,' 1828, vol. i. 165-6.

length cut, and all reasonable ground of complaint so far as respected the lands along the North Don was removed. For some years the new settlers cultivated their lands in peace; when suddenly they were reduced to the greatest distress, through the troubles arising out of the wars of the Commonwealth.

In 1642 a committee sat at Lincoln to watch over the interests of the Parliament in that county. The Yorkshire royalists were very active on the other side of the Don, and the rumour went abroad that Sir Ralph Humby was about to march into the Isle of Axholme with his forces. To prevent this, the committee at Lincoln gave orders to break the dykes, and pull up the flood-gates at Snow-sewer and Millerton-sluice. Thus in one night the results of many years' labour were undone, and the greater part of the level again lay under water. The damage inflicted on the Hatfield settlers in that one night was estimated at not less than 20,000*l.* The people who broke the dykes were, no doubt, glad to have the opportunity of taking their full revenge upon the foreigners for robbing them of their commons. They levelled the Dutchmen's houses, destroyed their growing corn, and broke down their fences; and, when some of them tried to stop the destruction of the sluices at Snow-sewer, the rioters stood by with loaded guns, and swore they would stay until the whole levels were drowned again, and the foreigners forced to swim away like ducks.

After the mischief had been done, the commoners set up their claims as participants in the lands which had not been drowned, from which the foreigners had been driven. In this they

I.   D

were countenanced by Colonel Lilburne, who, with
a force of Parliamentarians, occupied Sandtoft,
driving the Protestant minister out of his house,
and stabling their horses in his chapel. A bargain
was actually made between the Colonel and the
commoners, by which 2000 acres of Epworth
Common were to be assigned to him, on condition
of their right being established as to the remainder,
while he undertook to hold them harmless in respect of the cruelties which they had perpetrated
on the poor settlers of the level. When the injured parties attempted to obtain redress by law,
Lilburne, by his influence with the Parliament, the
army, and the magistrates, parried their efforts for
eleven years.* He was, however, eventually compelled to disgorge ; and though the original settlers
at length got a decree of the Council of State
in their favour, and those of them who survived
were again permitted to occupy their holdings, the
nature of the case rendered it impossible that
they should receive any adequate redress for their
losses and sufferings.†

In the mean time Sir Cornelius Vermuyden had

* Colonel Lilburne attempted an ineffectual defence of himself
in the tract entitled 'The Case of the Tenants of the Manor of
Epworth truly stated by Col. Jno. Lilburne,' Nov. 18th, 1651.

† For a long time after this, indeed, the commoners continued
at war with the settlers, and both were perpetually resorting to the
law—of the courts as well as of the strong hand. One Reading,
a counsellor, was engaged to defend the rights of the drainers or
participants, but his office proved a very dangerous one. The
fen-men regarded him as an enemy, and repeatedly endeavoured
to destroy him. Once they had nearly burned him and his family
in their beds. Reading died in 1716, at a hundred years old, fifty
of which he had passed in constant danger of personal violence,
having fought "thirty-one set battles" with the fen-men in defence
of the drainers' rights.

not been idle. He was as eagerly speculative as
ever. Before he parted with his interest in the
reclaimed lands at Hatfield, he was endeavouring
to set on foot his scheme for the reclamation of
the drowned lands in the Cambridge Fens; for we
find the Earl of Bedford, in July, 1630, writing to
Sir Harry Vane, recommending him to join Sir
Cornelius and himself in the enterprise. Before
the end of the year Vermuyden entered into a
contract with the Crown for the purchase of Malvern Chase, in the county of Worcester, for the
sum of 5000*l.*, which he forthwith proceeded to
reclaim and enclose. Shortly after he took a grant
of 4000 acres of waste land on Sedgemoor, with the
same object, for which he paid 12,000*l.* Then in
1631 we find him, in conjunction with Sir Robert
Heath, taking a lease for thirty years of the Dove-
gang lead-mine, near Wirksworth, reckoned the
best in the county of Derby. But from this point
he seems to have become involved in a series of
lawsuits, from which he never altogether shook
himself free. His connection with the Hatfield
estates got him into legal, if not pecuniary difficulties, and he appears for some time to have
suffered imprisonment. He was also harassed by
the disappointed Dutch capitalists at the Hague
and Amsterdam, who had suffered heavy losses
by their investments at Hatfield, and took legal
proceedings against him. He had no sooner,
however, emerged from confinement than we find
him fully occupied with his new and grand project
for the drainage of the Great Level of the Fens.

The outfalls of the numerous rivers flowing
through the Fen Level having become neglected,
the waters were everywhere regaining their old

dominion. Districts which had been partially reclaimed were again becoming drowned, and even
the older settled farms and villages situated upon
the islands of the Fens were threatened with like
ruin. The Commissioners of Sewers at Huntingdon attempted to raise funds for improving the
drainage by levying a tax of six shillings an acre
upon all marsh and fen lands, but not a shilling of
the tax was collected. This measure having failed,
the Commissioners of Sewers of Norfolk, at a
session held at King's Lynn, in 1629, determined
to call to their aid Sir Cornelius Vermuyden. At
an interview to which he was invited, he offered to
find the requisite funds to undertake the drainage
of the Level, and to carry out the works after the
plans submitted by him, on condition that 95,000
acres of the reclaimed lands were granted to him
as a recompense. A contract was entered into on
those terms; but so great an outcry was immediately raised against such an arrangement being
made with a foreigner, that it was abrogated before
many months had passed.

Then it was that Francis, Earl of Bedford, the
owner of many of the old church-lands in the Fens,
was induced to take the place of Vermuyden, and
become chief undertaker in the drainage of the
extensive tract of fen country now so well known
as the Great Bedford Level. Several of the adjoining landowners entered into the project with
the Earl, contributing sums towards the work, in
return for which a proportionate acreage of the
reclaimed lands was to be allotted to them. The
new undertakers, however, could not dispense
with the services of Vermuyden. He had, after
long study of the district, prepared elaborate plans

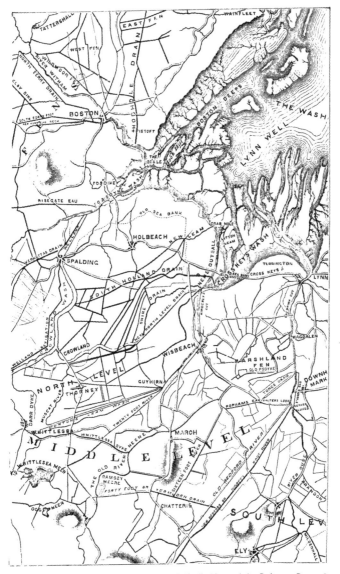

Map of the Fens as drained in 1830. [After Telford's Plan and the Ordnance Survey.]

for its drainage, and, besides, had at his command an organised staff of labourers, mostly Flemings, who were well accustomed to this kind of work. Westerdyke, also a Dutchman, prepared and submitted plans, but Vermuyden's were preferred, and he was accordingly authorised to proceed with the enterprise.

The difficulties encountered in carrying on the works were very great, arising principally from the want of funds. The Earl of Bedford became seriously crippled in his resources; he raised money upon his other property until he could raise no more, while many of the smaller undertakers were completely ruined. Vermuyden meanwhile took energetic measures to provide the requisite means to pay the workmen and prosecute the drainage; until the undertakers became so largely his debtors that they were under the necessity of conveying to him many thousand acres of the reclaimed lands, even before the works were completed, as security for the large sums which he had advanced.

The most important of the new works executed at this stage were as follows;—Bedford River (now known as Old Bedford River), extending from Erith on the Ouse to Salter's Lode on the same river: this cut was 70 feet wide and 21 miles long, and its object was to relieve and take off the high floods of the Ouse.* Bevill's Leam was another

* We insert the map on the preceding page at this place, although it includes the drainage-works subsequently constructed, in order that the reader may be enabled more readily to follow the history of the various cuts and drains executed in the Fen country from about the middle of the sixteenth century down to about the year 1830.

extensive cut, extending from Whittlesea Mere to
Guyhirne, 40 feet wide and 10 miles long; Sam's
Cut, from Feltwell to the Ouse, 20 feet wide and
6 miles long; Sandy's Cut, near Ely, 40 feet wide
and 2 miles long; Peakirk Drain, 17 feet wide and
10 miles long; with other drains, such as Milden-
hall, New South Eau, and Shire Drain. Sluices
were also erected at Tydd upon Shire Drain, at
Salter's Lode, and at the Horseshoe below Wisbeach, together with a clow,* at Clow's Cross, to
keep out the tides ; while a strong fresh-water
sluice was also provided at the upper end of the
Bedford River.

These works were not permitted to proceed
without great opposition on the part of the Fen-
men, who frequently assembled to fill up the cuts
which the labourers had dug, and to pull down
the banks which they had constructed. They also
abused and maltreated the foreigners when the
opportunity offered, and sometimes mobbed them
while employed upon the drains, so that in several
places they had to work under a guard of armed
men. Difficult though it was to deal with the
unreclaimed bogs, the unreclaimed "fen-slodgers"
were still more impracticable. Although their
condition was very miserable, they nevertheless
enjoyed a sort of wild liberty amidst the watery
wastes, which they were not disposed to give up.
Though they might alternately shiver and burn
with ague, and become prematurely bowed and
twisted with rheumatism, still the Fens were
their " native land," such as it was, and their only

---

* A clow is a sluice regulated by being lifted or dropped perpendicularly, like a portcullis. The other sluices open and shut
like gates.

source of subsistence, precarious though it might
be. The Fens were their commons, on which
their geese grazed. They furnished them with
food, though the finding thereof was full of adventure and hazard. What cared the Fen-men
for the drowning of the land? Did not the water
bring them fish, and the fish attract wild fowl,
which they could snare and shoot? Thus the
proposal to drain the Fens and to convert them
into wholesome and fruitful lands, however important in a national point of view, as enlarging the
resources and increasing the wealth of the country,
had no attraction whatever in the eyes of the Fen-
men. They muttered their discontent, and everywhere met the " adventurers," as the reclaimers were
called, with angry though ineffectual opposition.
But their numbers were too few, and they were
two widely scattered, to make any combined effort
at resistance. They could only retreat to other
fens where they thought they might still be safe,
carrying their discontent with them, and com-
plaining that their commons were taken from them
by the rich, and, what was worse, by foreigners—
Dutch and Flemings. The jealous John Bull of
the towns became alarmed at this idea, and had
rather that the water than these foreigners had
possession of the land. "What!" asked one of
the objectors, "is the old activitie and abilities of
the English nation grown now soe dull and insufficient that wee must pray in ayde of our neighbours
to improve our own demaynes? For matter of
securitie, shall wee esteem it of small moment to
put into the hands of strangers three or four such
ports as Linne, Wisbeach, Spalding, and Boston,
and permit the countrie within and between them

to be peopled with overthwart neighbours; or, if they quaile themselves, must wee give place to our most auncient and daungerous enemies, who will be readie enough to take advantage of soe manie fair inlets into the bosom of our land, lying soe near together that an army landing in each of them may easily meet and strongly entrench themselves with walls of water, and drowne the countrie about them at their pleasure?" *

Thus a great agitation against the drainage sprang up in the Fen districts, and a wide-spread discontent prevailed, which, as we shall afterwards find, exercised an important influence on the events which culminated in the Great Rebellion of a few years later. Among the other agencies brought to bear against the Fen drainers was the publication of satirical songs and ballads—the only popular press of the time; and the popular poets doubtless represented accurately enough the then state of public opinion, as their ballads were sung with great applause about the streets of the Fen towns. One of these, entitled, 'The Powte's† Complaint,' was among the most popular.

* 'The Drayner Confirmed,' tract, 1629.
† Powte—the old English word for the sea-lamprey.

THE POWTE'S COMPLAINT.

Come, Brethren of the water, and let us all assemble,
To treat upon this Matter, which makes us quake and tremble;
For we shall rue, if it be true that Fens be undertaken,
And where we feed in Fen and Reed, they'll feed both Beef and Bacon.

They'll sow both Beans and Oats, where never Man yet thought it;
Where Men did row in Boats, ere Undertakers bought it;
But, *Ceres*, thou behold us now, let wild Oats be their Venture,
Oh, let the Frogs and miry Bogs destroy where they do enter.

Behold the great Design, which they do now determine,
Will make our Bodies pine, a prey to Crows and Vermine ·

In another popular drinking song, entitled 'The Draining of the Fennes,' the Dutchmen are pointed out as the great offenders. The following stanzas may serve as a specimen : —

> " The Dutchman hath a thirsty soul,
>   Our cellars are subject to his call ;
> Let every man, then, lay hold on his bowl,
>   'Tis pity the German sea should have all.

For they do mean all Fens to drain, and Waters overmaster,
All will be dry, and we must die—'cause Essex Calves want pasture.

Away with Boats and Rudder, farewel both Boots and Skatches,
No need of one nor t'other, Men now make better Matches ;
Stilt-makers all, and Tanners, shall complain of this Disaster,
For they will make each muddy lake for Essex Calves a Pasture.

The feather'd Fowls have Wings, to fly to other Nations ;
But we have no such things, to help our Transportations ;
We must give place, O grevious Case ! to horned Beasts and Cattle
Except that we can all agree to drive them out by Battel.

Wherefore let us intreat our antient Water-Nurses
To shew their Power so great as t' help to drain their Purses
And send us good old Captain Flood to lead us out to Battel,
Then Two-penny Jack, with Scales on 's Back, will drive out all the Cattle.

This Noble Captain yet was never known to fail us,
But did the conquest get of all that did assail us ;
His furious Rage none could assuage ; but, to the World's great Wonder
He tears down Banks, and breaks their Cranks and Whirligigs assunder.

God *Eolus*, we do thee pray that thou wilt not be wanting ;
Thou never said'st us nay—now listen to our canting ;
Do thou deride their Hope and Pride that purpose our Confusion,
And send a Blast that they in haste may work no good Conclusion.

Great *Neptune*, God of Seas, this Work must needs provoke ye ;
They mean thee to disease, and with Fen-Water choak thee ;
But with thy Mace do thou deface, and quite confound this matter,
And send thy Sands to make dry lands when they shall want fresh Water.

And eke we pray thee, *Moon*, that thou wilt be propitious,
To see that nought be done to prosper the Malicious ;
Tho' Summer's Heat hath wrought a Feat, whereby themselves they flatter,
Yet be so good as send a Flood, lest Essex Calves want Water.

> Then apace, apace drink, drink deep, drink deep,
> Whilst 'tis to be had let's the liquor ply ;
> The drainers are up, and a coile they keep,
> And threaten to drain the kingdom dry.
>
> Why should we stay here, and perish with thirst ;
> To th' new world in the moon away let us goe,
> For if the Dutch colony get thither first,
>   'Tis a thousand to one but they'll drain that too !
>   *Chorus*— Then apace, apace drink, &c."

The Fen drainers might, however, have out-lived these attacks, had the works executed by them been successful ; but unhappily they failed in many respects. Notwithstanding the numerous deep cuts made across the Fens in all directions at such great cost, the waters still retained their hold upon the land. The Bedford River and the other drains merely acted as so many additional receptacles for the surplus water, without relieving the drowned districts to any appreciable extent. This arose from the engineer confining his attention almost exclusively to the inland draining and embankments, while he neglected to provide any sufficient outfalls for the waters themselves into the sea. Vermuyden committed the error of adopting the Dutch method of drainage, in a district where the circumstances differed in many material respects from those which prevailed in Holland. In Zeeland, for instance, the few rivers passing through it were easily banked up and carried out to sea, whilst the low-lying lands were kept clear of surplus water by pumps driven by windmills. There, the main object of the engineer was to build back the river and the ocean ; whereas in the Great Level the problem to be solved was, how to provide a ready outfall to the sea for the vast body of fresh water falling upon as well as flowing through the Fens themselves. This essential point was unhappily overlooked by the early drainers; and it has thus happened that the chief work of modern engineers has been to rectify the errors of Vermuyden and his followers; more especially by providing efficient outlets for the discharge of the Fen waters, deepening and straightening the rivers, and compressing the streams in their course through the Level, so as to produce a more powerful current and scour, down to their point of outfall into the sea.

This important condition of successful drainage having been overlooked, it may readily be understood how unsatisfactory was the result of the works first carried out in the Bedford Level. In some districts the lands were no doubt improved by the additional receptacles provided for the surplus waters, but the great extent of fen land still lay for the most part wet, waste, and unprofitable. Hence, in 1634, a Commission of Sewers held at Huntingdon pronounced the drainage to be defective, and the 400,000 acres of the Great Level to be still subject to inundation, especially in the winter season. The King, Charles I., then resolved himself to undertake the reclamation, with the object of converting the Level, if possible, into "winter grounds." He took so much personal interest in the work that he even designed a town to be called Charleville, which was to be built in the midst of the Level, for the purpose of commemorating the undertaking. Sir Cornelius Vermuyden was again employed, and he proceeded to carry out the King's design. He had many enemies, but he could not be dispensed with; being the only man of recognised ability in works of drainage at that time in England.

The works constructed in pursuance of this new design were these:—an embankment on the south side of Morton's Leam, from Peterborough to Wisbeach; a navigable sasse, or sluice, at Standground; a new river cut between the stone sluice at the Horse-shoe and the sea below Wisbeach, 60 feet broad and 2 miles long, embanked at both sides; and a new sluice in the marshes below Tydd, upon the outfall of Shire Drain. These and other works were in full progress, when the political troubles of the time came to a height, and brought all operations to a standstill for many years. The discontent caused throughout the Fens by the drainage operations had by no means abated; but, on the contrary, considerably increased. In other parts of the kingdom, the attempts made about the same time by Charles I. to levy taxes without the authority of Parliament gave rise to much agitation. In 1637 occurred Hampden's trial, arising out of his resistance to the payment of ship-money: by the end of the same year the King and the Parliamentary party were mustering their respective forces, and a collision between them seemed imminent.

At this juncture the discontent which prevailed throughout the Fen counties was an element of influence not to be neglected. It was adroitly represented that the King's sole object in draining the Fens was merely to fill his impoverished exchequer, and enable him to govern without a Parliament. The discontent became fanned into a fierce flame; on which Oliver Cromwell, the

member for Huntingdon, until then comparatively
unknown, availing himself of the opportunity
which offered, of increasing the influence of the
Parliamentary party in the Fen counties, immediately put himself at the head of a vigorous
agitation against the further prosecution of the
scheme. He was very soon the most popular
man in the district; he was hailed 'Lord of the
Fens' by the Fen-men; and he went from meeting to meeting, stirring up the public discontent,
and giving it a suitable direction. "From that
instant," says Mr. Forster,* "the scheme became
thoroughly hopeless. With such desperate determination he followed up his purpose—so actively
traversed the district, and inflamed the people
everywhere—so passionately described the greedy
claims of royalty, the gross exactions of the commission, nay, the questionable character of the
improvement itself, even could it have gone on
unaccompanied by incidents of tyranny,—to the
small proprietors insisting that their poor claims
would be merely scorned in the new distribution of the property reclaimed,—to the labouring
peasants that all the profit and amusement they
had derived from *commoning* in those extensive
wastes were about to be snatched for ever from
them,—that, before his almost individual energy,
King, commissioners, noblemen-projectors, all were
forced to retire, and the great project, even in the
state it then was, fell to the ground."

The success of the Cambridge Fen-men, in
resisting the reclamation of the wastes, encouraged
those in the more northern districts to take even

* Lives of Eminent British Statesmen (Lardner's 'Cabinet
Cyclopædia,' vol. vi. p. 60).

more summary measures to get rid of the drainers, and restore the lands to their former state. The Earl of Lindsey had succeeded at great cost in enclosing and draining about 35,000 acres of the Lindsey Level, and induced numerous farmers and labourers to settle upon the land. They erected dwellings and farm-buildings, and were busily at work, when the Fen-men suddenly broke in upon them, destroyed their buildings, killed their cattle, and let in the waters again upon the land. So, too, in the West and Wildmore Fen district between Tattershall and Boston in Lincolnshire, where considerable progress had been made by a body of "adventurers" in reclaiming the wastes. After many years' labour and much cost, they had succeeded in draining, enclosing, and cultivating an extensive tract of rich land, and they were peaceably occupied with their farming pursuits, when a mob of Fen-men collected from the surrounding districts, and under pretence of playing at football, levelled the enclosures, burnt the corn and the houses, destroyed the cattle, and even killed many of the people who occupied the land. They then proceeded to destroy the drainage works, by cutting across the embankments and damming up the drains, by which the country was again inundated and restored to its original state.

The greater part of the Level thus again lay waste, and the waters were everywhere extending their dominion over the dry land through the choking up of the drains and river outfalls by the deposit of silt. Matters were becoming even worse than before, but could not be allowed thus to continue. In 1641 the Earl of Bedford and his participants made an application to the Long

Parliament, then sitting, for permission to re-enter upon the works; but the civil commotions which still continued prevented any steps being taken, and the Earl himself shortly after died in a state of comparative penury, to which he had reduced himself by his devotion to this great work. Again, however, we find Sir Cornelius Vermuyden upon the scene. Undaunted by adversity, and undismayed by the popular outrages committed upon his poor countrymen in Lincolnshire and Yorkshire, he still urged that the common weal of England demanded that the rich lands lying under the waters of the Fens should be reclaimed, and made profitable for human uses. He saw a district almost as large as the whole of the Dutch United Provinces remaining waste and worse than useless, and he gave himself no rest until he had set on foot some efficient measure for its drainage and reclamation. What part he took in the political discussions of the time, we know not; but we find the eldest of his sons, Cornelius, a colonel in the Parliamentary army * stationed in the Fens under Fairfax, shortly before the battle of Naseby. Vermuyden himself was probably too much engrossed by his drainage project to give heed to political affairs; and besides, he could not forget that Charles, and Charles's father, had been his fast friends.

* "The party under Vermuyden waits the King's army, and is about Deeping ; has a command to join with Sir John Gell, if he commands him."—Cromwell's Letter to Fairfax, 4th June, 1645. This Vermuyden resigned his commission a few days before the battle of Naseby, having, as he alleged, special reasons requiring his presence beyond the seas, whence he does not seem to have returned until after the Restoration. In 1665 we find him a member of the Corporation of the Bedford Level.

LOCKS MILL (AT SPALDING) IN THE FEN COUNTRY.

In 1642, while the civil war was still raging, appeared Vermuyden's 'Discourse' on the Drainage of the Fens, wherein he pointed out the works which still remained to be executed in order effectually to reclaim the 400,000 acres of land capable of growing corn, which formed the area of the Great Level. His suggestions formed the subject of much pamphleteering discussion for several years, during which also numerous petitions were presented to Parliament, urging the necessity for perfecting the drainage. At length, in 1649, authority was granted to William, Earl of Bedford, and other participants, to prosecute the undertaking which his father had begun, and steps were shortly after taken to recommence the works. Again was Westerdyke, the Dutch engineer, called in to criticise Vermuyden's plans; and again was Vermuyden triumphant over his opponent. He was selected, once more, to direct the drainage, which, looking at the defects of the works previously executed by him, and the difficulties in which the first Earl had thereby become involved, must be regarded as a marked proof of the man's force of purpose, as well as of his recognised integrity of character.

Vermuyden again collected his Dutchmen about him, and vigorously began operations. But they had not proceeded far before they were again almost at a standstill for want of funds; and throughout their entire progress they were hampered and hindered by the same great difficulty. Some of the participants sold and alienated their shares in order to get rid of further liabilities; others held on, but became reduced to the lowest ebb. Means were, however, adopted to obtain a

supply of cheaper labour; and application was
made by the adventurers for a supply of men
from amongst the Scotch prisoners who had been
taken at the battle of Dunbar. A thousand of
them were granted for the purpose, and employed
on the works to the north of Bedford River,
where they continued to labour until the political
arrangements between the two countries enabled
them to return home. When the Scotch labourers
had left, some difficulty was again experienced in
carrying on the works. The local population
were still hostile, and occasionally interrupted the
labourers employed upon them; a serious riot at
Swaffham having only been put down by the
help of the military. Blake's victory over Van
Tromp, in 1652, opportunely supplied the Government with a large number of Dutch prisoners, five
hundred of whom were at once forwarded to the
Level, where they proved of essential service as
labourers.

The most important of the new rivers, drains,
and sluices included in this further undertaking,
were the following:—The New Bedford River,
cut from Erith on the Ouse to Salter's Lode on
the same river, reducing its course between these
points from 40 to 20 miles: this new river was
100 feet broad, and ran nearly parallel with the
Old Bedford River. A high bank was raised
along the south side of the new cut, and an
equally high bank along the north side of the old
river, a large space of land, of about 5000 acres,
being left between them, called the Washes, for
the floods to "bed in," as Vermuyden termed it.
Then the river Welland was defended by a bank,
70 feet broad and 8 feet high, extending from

Peakirk to the Holland bank. The river Nene was also defended by a similar bank, extending from Peterborough to Guyhirne; and another bank was raised between Standground and Guyhirne, so as to defend the Middle Level from the overflowing of the Northamptonshire waters. The river Ouse was in like manner restrained by high banks extending from Over to Erith, where a navigable sluice was provided. Smith's Leam was cut, by which the navigation from Wisbeach to Peterborough was opened out. Among the other cuts and drains completed at the same time, were Vermuyden's Eau, or the Forty Feet Drain, extending from Welch's Dam to the river Nene near Ramsey Mere; Hammond's Eau, near Somersham, in the county of Huntingdon; Stonea Drain and Moore's Drain, near March, in the Isle of Ely; ·Thurlow's Drain, extending from the Forty Feet to Popham's Eau; and Conquest Lode, leading to Whittlesea Mere. And in order to turn the tidal waters into the Hundred Feet River, as well as to prevent the upland floods from passing up the Ten Mile River towards Littleport, Denver Sluice, that great bone of after contention, was constructed. Another important work in the South Level was the cutting of a large river called St. John's, or Downham Eau,* 120 feet wide, and 10 feet deep, from Denver Sluice to Stow Bridge on the Ouse, with sluices at both ends, for the purpose of carrying away with greater facility the flood waters descending from the several rivers of that level. Various new sluices were also fixed at the mouths

---

* The St. John's Eau, being a straight cut, is known in the district as "The Poker;" and Marshland Cut, being in the shape of a pair of tongs, is commonly called "Tong Drain."

of the rivers, to prevent the influx of the tides, and most of the old drains and cuts were at the same time scoured out and opened for the more ready flow of the surface waters.

At length, in March, 1652, the works were declared to be complete, and the Lords Commissioners of Adjudication appointed under the Act of Parliament proceeded to inspect them. They embarked upon the New River, and sailing over it to Stow Bridge, surveyed the new eaus and sluices executed near that place, after which they returned to Ely. There Sir Cornelius read to those assembled a discourse, in which he explained the design he had carried out for the drainage of the district; in the course of which he stated as one of the results of the undertaking, that in the North and Middle Levels there were already 40,000 acres of land " sown with cole-seed, wheat, and other winter grain, besides innumerable quantities of sheep, cattle, and other stock, where never had been any before. " These works," he added, " have proved themselves sufficient, as well by the great tide about a month since, which overflowed Marsh-land banks, and drowned much ground in Lincolnshire and other places, and a flood by reason of a great snow, and rain upon it following soon after, and yet never hurt any part of the whole Level; and the view of them, and the consideration of what hath previously been said, proves a clear draining according to the Act." He concluded thus,—"I presume to say no more of the work, lest I should be accounted vain-glorious; although I might truly affirm that the present or former age have done nothing like it for the general good of the nation. I humbly desire that God may have the glory, for his blessing and bringing to perfection my poor endeavours, at the vast charge of the Earl of Bedford and his participants."

A public thanksgiving took place to celebrate the completion of the undertaking; and on the 27th of March, 1653, the Lords Commissioners of Adjudication of the Reclaimed Lands, accompanied by their officers and suite,—the Company of Adventurers, headed by the Earl of Bedford,—the magistrates and leading men of the district, with a vast concourse of other persons, — attended public worship in the cathedral of Ely, when the Rev. Hugh Peters, chaplain to the Lord-General Cromwell, preached a sermon on the occasion.

Vermuyden's perseverance had thus far triumphed. He had stood by his scheme when all others held aloof from it. Amidst the engrossing excitement of the civil war, the one dominating idea which possessed him was the drainage of the Great Level. While the nation was divided into two hostile camps, and the deadly struggle was proceeding between the Royalists and the Parliamentarians, Vermuyden's sole concern was how to raise the funds wherewith to pay his peaceful army of Dutch labourers in the Fens. To carry on the works he sold every acre of the soil he had reclaimed. He first sold the allotment of land won by him from the Thames at Dagenham in 1621; then he sold his interest in his lands at Sedgemoor and Malvern Chase; and in 1654 we find him conveying the remainder of his property in Hatfield Level. He was also under the necessity of selling all the lands apportioned to him in the Bedford Level itself, in order to pay the debts incurred in their drainage. But although he lost all, it appears that the company

in the end preferred heavy pecuniary claims
against him which he had no means of meeting;
and in 1656 we find him appearing before Parliament as a suppliant for redress. Thenceforward
he entirely disappears from public sight; and it is
supposed that, very shortly after, he went abroad
and died, a poor, broken down old man, the extensive lands which he had reclaimed and owned
having been conveyed to strangers.

The drainage of the Fens, however, was not yet
complete. The district was no longer a boggy
wilderness, but much of it in fine seasons was
covered with waving crops of corn. As the swamps
were drained, farm buildings, villages, and towns
gradually sprang up, and the toil of the labourer
was repaid by abundant harvests. The anticipation
held forth in the original charter granted by
Charles I. to the reclaimers of the Bedford Level
was more than fulfilled. "In those places which
lately presented nothing to the eyes of the be-
holders but great waters and a few reeds thinly
scattered here and there, under the Divine mercy
might be seen pleasant pastures of cattle and kine,
and many houses belonging to the inhabitants."
But the tenure by which the land continued to be
held was unremitting vigilance * and industry; the

---

* Since the above passage was written, in 1861, its truth has
been amply confirmed by the blowing up of the Middle Level sluice,
about three miles above Lynn, by which some 10,000 acres of the
richest agricultural land in Marshland were completely submerged.
Great loss was thereby occasioned, and a fertile crop of lawsuits
has followed the inundation. The Middle Level Drain was admirably planned by the late James Walker, C.F., and was supposed
to have been as admirably executed, at a cost of over 400,000*l.*
For ten years it was a complete success; but the foundations of
the outfall sluice having become undermined by the tidal scour in

difficulties interposed by nature tending to discipline the skill, to stimulate the enterprise, and evoke the energy of the people who had rescued the fields from the watery waste.

Improvements of all kinds went steadily on, until all the rivers flowing through the Level were artificially banked and diverted into new channels, excepting the Nene, which is the only natural river in the Fen district remaining comparatively unaltered. New dykes, causeways, embankments, and sluices were formed; many droves, leams, eaus, and drains were cut, furnished with gowts or gates at their lower ends, which were from time to time dug, deepened, and widened. Mills were set to work to pump out the water from the low grounds; first windmills, sometimes with double-lifts, as practised in Holland; and more recently powerful steam-engines. Sluices were also erected to prevent the inland waters from returning; strong embankments extending in all directions, to keep the rivers and tides within their defined channels. To protect the land from the sea waters as well as the fresh,—to build and lock back the former, and to keep the latter within due limits,—was the work of the engineer; and by his skill, aided by the industry of his contractors and workmen, water, instead of being the master and tyrant as of old, became man's servant and pliant agent, and was used as an irrigator, a conduit, a mill-stream, or a water-road for extensive districts of country. In short, in no part of the world, except

the Ouse, the masonry fell in, and the waters immediately rushed in upon the land and resumed possession. The drainage has since been restored in a very able and efficient manner by Mr. Hawkshaw, and Marshland is again under cultivation.

in Holland, have more industry and skill been displayed in reclaiming and preserving the soil, than in Lincolnshire and the districts of the Great Bedford Level. Six hundred and eighty thousand acres of the most fertile land in England, or an area equal to that of North and South Holland, have been converted from a dreary waste into a fruitful plain, and fleets of vessels traverse the district itself, freighted with its rich produce. Taking its average annual value at 4*l.* an acre, the addition to the national wealth and resources may be readily calculated.

The prophecies of the decay that would fall upon the country, if "the valuable race of Fen-men" were deprived of their pools for pike, and fish, and wild-fowl, have long since been exploded. The population has grown in numbers, in health, and in comfort, with the progress of drainage and reclamation. The Fens are no longer the lurking places of disease,* but as salubrious as any other parts of England. Dreary swamps are supplanted by pleasant pastures, and the haunts of pike and wild-fowl have become the habitations of industrious farmers and husbandmen. Even Whittlesea Mere and Ramsey Mere,—the only two lakes, as we were told in the geography books

---

* When Dr. Whalley was presented by the Bishop of Ely to the rectory of Hagworthingham-in-the-Fens, it was with the singular proviso that he was not to reside in it, as the air was fatal to any but a native. (*Journals and Correspondence of T. S. Whalley, D.D.*) Statistics, however, prove that the Fen districts are the exceptional haunts of disease. The Registrar-General, in one of his recent reports, states that, whilst the mortality of Pau in the Pyrenees, a place resorted to by British invalids on account of its salubriousness, is 23 in 1000, that of Ely is only 17 in 1000.

of our younger days, to be found in the south of England,—have been blotted out of the map, for they have been drained by the engineer, and are now covered with smiling farms and pleasant homesteads.

CHAPTER III.

SIR HUGH MYDDELTON.—THE CUTTING OF THE NEW RIVER.

WHILE the engineer has occasionally to contend with all his skill against the powers of water, he has also to deal with it as a useful agent. Though water, like fire, is a bad master, the engineer contrives to render it docile and tractable. He leads it in artificial channels for the purpose of driving mills and machinery, or he employs it to feed canals along which boats and ships laden with merchandise may be safely floated.

But water is also an indispensable necessary of life, an abundant supply of it being essential for human health and comfort. Hence nearly all the ancient towns and cities were planted by the banks of rivers, principally because the inhabitants required a plentiful supply of water for their daily uses. Old London had not only the advantage of its pure broad stream flowing along its southern boundary, so useful as a water-road, but it also possessed an abundance of Wells, from which a supply of pure water was obtained, adequate for the requirements of its early population. The river of Wells, or Wallbrook, flowed through the middle of the city ; and there were numerous wells in other quarters, the chief of which were Clerke's Well, Clement's Well, and Holy Well, the names of which still survive in the streets built over them.

As London grew in size and population, these wells were found altogether inadequate for the wants of the inhabitants; besides, the water drawn from them became tainted by the impurities which

Ancient Conduit in Westcheap.

filter into the soil wherever large numbers are congregated. Conduits were then constructed, through which water was led from Paddington, from James's Head, Mewsgate, Tyburn, Highbury and Hampstead. There were sixteen of such public conduits about London, and the Conduit Streets which still exist throughout the metropolis mark the sites of

several of these ancient works.* The copious
supply of water by the conduits was all the more
necessary at that time, as London was for the most
part built of timber, and liable to frequent fires, to
extinguish which promptly, every citizen was bound
to have a barrel full of water in readiness outside
his door. The corporation watched very carefully
over their protection, and inflicted severe punish-
ments on such as interfered with the flow of water
through them. We find a curious instance of this
in the City Records, from which it appears that,
on the 12th November, 1478, one William Campion,
resident in Fleet Street, had cunningly tapped the
conduit where it passed his door, and conveyed
the water into a well in his own house, "thereby
occasioning a lack of water to the inhabitants."
Campion was immediately had up before the Lord
Mayor and Aldermen, and after being confined for
a time in the Comptour in Bread Street, the following further punishment was inflicted on him. He
was set upon a horse with a vessel like unto a

---

* The conduits used, in former times, to be yearly visited with
considerable ceremony. For instance, we find that—"On the 18th
of September, 1562, the Lord Mayor (Harper), the Aldermen, with
many worshipful persons, and divers of the Masters and Wardens
of the twelve companies, rode to the Conduit's-head [now the site
of Conduit Street, New Bond Street], for to see them after the old
custom. And afore dinner they hunted the hare and killed her,
and thence to dinner at the head of the Conduit. There was a
good number entertained with good cheere by the Chamberlain,
and, after dinner, they hunted the fox. There was a great cry for
a mile, and at length the hounds killed him at the end of St. Giles's.
Great hallooing at his death, and blowing of hornes ; and thence
the Lord Mayor, with all his company, rode through London to his
place in Lombard Street." Stow's 'Survey of London.'—It would
appear that the ladies of the Lord Mayor and Aldermen attended
on these jovial occasions, riding *in waggons*.

conduit placed upon his head, which being filled with water running out of small pipes from the same vessel, he was taken round all the conduits of the city, and the Lord Mayor's proclamation of his offence and the reason for his punishment was then read. When the conduit had run itself empty over the culprit, it was filled again. The places at which the proclamation was read were the following,—at Leadenhall, at the pillory in Cornhill, at the great conduit in Chepe, at the little conduit in the same street, at Ludgate and Fleet Bridge, at the Standard in Fleet Street, at Temple Bar, and at St. Dunstan's Church in Fleet Street; from whence he was finally marched back to the Comptour, there to abide the will of the Lord Mayor and Aldermen.*

But the springs from which the conduits were supplied in course of time decayed; perhaps they gradually diminished by reason of the sinking of wells in their neighbourhood for the supply of the increasing suburban population. Hence a deficiency of water began to be experienced in the city, which in certain seasons almost amounted to a famine. There were frequent contentions at the conduits for "first turn," and when water was scarce, these sometimes grew into riots. The water carriers came prepared for a fight, and at length the Lord Mayor had to interfere, and issued his proclamation forbidding persons from resorting to the conduits armed with clubs and staves. This, however, did not remedy the deficiency. It is true the Thames,—"that most delicate and serviceable river," as Nichols terms it,† was always

* 'Corporation Records,' Index No. I., fo. 184 b.

† 'Progresses of James I.,' vol. ii., 699. The Corporation records contain numerous references to the preservation of the

available; but an increasing proportion of the inhabitants lived at a distance from the river.* Besides, the attempt was made by those who occupied the lanes leading towards the Thames to stop the thoroughfare, and allow none to pass without paying a toll. A large number of persons then obtained a living as water carriers,* selling the water by the "tankard" of about three gallons; and they seem to have formed a rather unruly portion of the population.

The difficulty of supplying a sufficient quantity of water to the inhabitants by means of wells, conduits, and water carriers, continued to increase, until the year 1582, when Peter Morice, the Dutchman, undertook, as the inhabitants could not go to the Thames for their water, to carry the Thames to them. With this object he erected an ingenious pumping engine in the first arch of London Bridge, worked by water wheels driven by the rise and fall of the tide, which then rushed with great velocity through the arches. This machine forced the water through leaden pipes, laid into the houses of the citizens. The power with which Morice's

purity of the water in the river. The Thames also furnished a large portion of the food of the city, and then abounded in salmon and other fish, the London fishermen constituting a large class. We find numerous proclamations issued relative to the netting of the "salmon and porpoises," wide nets and wall nets being especially prohibited. Fleets of swans on the Thames were a picturesque feature of the river down even to the time of James II.

* The water carrier was commonly called a "Cob," and Ben Jonson seems to have given a sort of celebrity to the character by his delineation of "Cob" in his 'Every Man in his Humour.' Gifford, in a note on the play, pointed out that there is an avenue still called "Cob's Court," in Broadway, Blackfriars; not improbably (he adds) from its having formerly been inhabited principally by the class of water carriers.

forcing pumps worked was such, that he was enabled to throw the water over St. Magnus's steeple,
greatly to the astonishment of the Mayor and
Aldermen, who assembled to witness the experiment. The machinery succeeded so well that a
few years later we find the corporation empowering the same engineer to use the second arch of
London Bridge for a similar purpose.*

But even this augmented machinery for pumping was found inadequate for the supply of London.
The town was extending rapidly in all directions,
and the growing density of the population along the
river banks was every year adding to the impurity
of the water, and rendering it less and less fit for
domestic purposes. Hence the demand for a more
copious and ready supply of pure water continued
steadily to increase. Where was the new supply
to be obtained, and how was it to be rendered
most readily available for the uses of the citizens?
Water is by no means a scarce element in England ;
and no difficulty was experienced in finding a sufficiency of springs and rivers of pure water at no
great distance from the metropolis. Thus, various
springs were known to exist in different parts of
Hertfordshire and Middlesex ; and many vague projects were proposed for conveying their waters to
London.

Desiring that one plan or another might be
carried out, the corporation obtained an Act
towards the end of Queen Elizabeth's reign,†

---

* The river pumping-leases continued in the family of the
Morices until 1701, when the then owner sold his rights to Richard
Soams for 38,000*l.*, and by him they were afterwards transferred to
the New River Company at a still higher price.

† Act 13 Eliz. c. 18.

empowering them to cut a river to the city from
any part of Middlesex or Hertfordshire; and ten
years were specified as the time for carrying out
the necessary works. But, though many plans
were suggested and discussed, no steps were taken
to cut the proposed river. The enterprise seemed
too large for any private individual to undertake;
and though the corporation were willing to sanction it, they were not disposed to find any part of
the requisite means for carrying it out. Notwithstanding, therefore, the necessity for a large supply
of water, which became more urgent in proportion
to the increase of population, the powers of the Act
were allowed to expire without anything having
been done to carry them into effect.

In order, however, to keep alive the parliamentary powers, another Act was obtained in the
third year of James I.'s reign (1605),* to bring an
artificial stream of pure water from the springs
of Chadwell and Amwell, in Hertfordshire; and
the provisions of this Act were enlarged and
amended in the following session.† From an entry
in the journals of the corporation, dated the 14th
October, 1606, it appears that one William Ingle-
bert petitioned the court for liberty to bring
the water from the above springs to the northern
parts of the city "in a trench or trenches of brick."
The petition was "referred," but nothing further
came of it; and the inhabitants of London continued for some time longer to suffer from the
famine of water—the citizens patiently waiting for
the corporation to move, and the corporation as
patiently waiting for the citizens.

The same inconveniences of defective water

* 3 Jac. c. 18.   † 4 Jac. c. 12.

supply were experienced in other towns, and measures were in some cases taken to remedy them. Thus, at Hull, in certain seasons, the inhabitants were under the necessity of bringing the water required by them for ordinary uses across the Humber from Lincolnshire in boats, at great labour and expense. They sought to obtain a better supply by leading water into the town from the streams in the neighbourhood; but the villagers of Hessle, Anlaby, and Cottingham, with others, resisted their attempts. In 1376, the mayor and burgesses appealed to the Crown; commissioners were appointed to inquire into the subject; and the result was, that powers were granted for making an aqueduct from Anlaby Springs to Hull. This was not accomplished without serious opposition on the part of the villagers, who riotously assembled to destroy the works, and they even went so far as to threaten Hull itself with destruction. Some of the rioters were seized and hanged at York, and the aqueduct was then finished.

Tiverton, in Devonshire, has in like manner been supplied with water from a very early period, by means of an artificial cut called the Town-leet, extending from a spring on White Down, about five miles distant, into the heart of the town. This valuable conduit was the free gift of Amicia, Countess of Devon, to the inhabitants, as long ago as the year 1240; and it continues a constant source of blessing. A perambulation is made along its course once in every five or six years, by the portreeve, the steward of the manor, the water bailiffs, and others, from Cogan's Well in the centre of the town, to the source of the

stream on White Down. All obstructions are
then removed, and the stream is claimed publicly
for the sole use of the inhabitants of Tiverton.
For about two miles of its course, it may, perhaps, be regarded as a natural stream; but from
the village of Chettiscombe the channel is for the
most part artificial, the water being confined within a high embankment, in many places above the
level of the surrounding country; and it is conveyed, as one writer says, "over a deep road behind the hospital, by a leaded shute, on a strong
stone arch, into the town." *   By this channel,
the water, shedding in its passage an allotted
portion to each street, is brought to Cogan's Well,
where it is artificially parted into three streams,
which run along the sides of the remaining streets,
until they are discharged into one or other of the
two rivers, the Loman and the Exe, from which
the place derives its name of Twy-ferd-town, or
Tiverton.   Copious rivulets are in like manner
led, by artificial cuts, through the principal streets
of Salisbury, from the natural streams at the confluence of which that city is situated.

But the most important artificial work of the
kind in the West of England is that constructed
for the supply of water to Plymouth, which was
carried out through the public spirit and enterprise of one of the most distinguished of English
admirals — the great Sir Francis Drake. It appears from ancient records that water was exceedingly scarce in that town, the inhabitants being
under the necessity of sending their clothes more
than a mile to be washed, the water used by them
for domestic purposes having to be fetched for the

* Dunsford's 'Historical Memoirs,' p. 106.

most part from Plympton, about five miles distant.
Sir Francis Drake, who was born within ten miles
of Plymouth, and had settled in the neighbourhood,
after having realised a considerable fortune by his
adventures on the Spanish main, observing the
great inconvenience suffered by the population
from their want of water, as well as the difficulty
of furnishing a supply to the ships frequenting
the port, conceived the project of remedying the
defect by leading a store of water to the town
from one of the numerous springs on Dartmoor.
Accordingly, in 1587, when he represented Bos-
siney (Tintagel) in Cornwall, he obtained an Act
enabling him to convey. a stream from the river
Mew or Meavy; and in the preamble to the Act
it was expressed that its object was not only to
ensure a continual supply of water to the inhabitants, but to obviate the inconvenience hitherto
sustained by seamen in watering their vessels.   It
would appear, from documents still extant, that
the town of Plymouth contributed 200*l.* towards
the expenses of the works, Sir Francis being at the
remainder of the cost ; and on the completion of
the undertaking the corporation agreed to grant
him a lease of the aqueduct for a term of twenty
years, at a nominal rental.   Drake lost no time in
carrying out the work, which was finished in four
years after the passing of the Act ; and its completion in 1591, on the occasion of the welcoming
of the stream into the town, was celebrated by
great public rejoicings.*

The " Leet," as it is called, is a work of no great

---

* The tradition survives to this day that Sir Francis Drake did
not cut the Leet by the power of money and engineering skill, but
by the power of magic.   It is said of him that, calling for his horse,

magnitude, though of much utility. It was originally nothing more than an open trench cut along
the sides of the moor, in which the water flowed
by a gentle inclination into the town and through
the streets of Plymouth. The distance between
the head of the aqueduct at Sheep's Tor and Plymouth, as the crow flies, is only seven miles;
but the length of the Leet—so circuitous are its
windings — is nearly twenty-four miles. After
its completion, Drake presented the aqueduct to
the inhabitants of Plymouth "as a free gift for
ever," and it has since remained vested in the
corporation, — who might, however, bestow more
care than they do on its preservation against impurity. Two years after the completion of the
Leet, the burgesses, probably as a mark of their
gratitude, elected Drake their representative in
Parliament. The water proved of immense public
convenience, and Plymouth, instead of being one
of the worst supplied, was rendered one of the
best watered towns in the kingdom. Until a
comparatively recent date the water flowed from
various public conduits, and it ran freely on either
side of the streets, that all classes of the people
might enjoy the benefit of a full and permanent
supply throughout the year. One of the original
conduits still remains at the head of Old Town-
street, bearing the inscription, "Sir Francis Drake
first brought this water into Plymouth, 1591."
    The example of Plymouth may possibly have

he mounted it and rode about Dartmoor until he came to a spring
sufficiently copious for his design, on which, pronouncing some
magical words, he wheeled round, and, starting off at a gallop, the
stream formed its own channel, and followed his horse's heels into
the town.

had an influence upon the corporation of London in obtaining the requisite powers from Parliament to enable them to bring the springs of Chadwell and Amwell to the thirsty population of the metropolis; but unhappily they had as yet no Drake to supply the requisite capital and energy. In March, 1608, one Captain Edmond Colthurst petitioned the Court of Aldermen for permission to enter upon the work; but it turned out that the probable cost was far beyond the petitioner's means, without the pecuniary help of the corporation; and that being withheld, the project fell to the ground. After this, one Edward Wright is said to have actually begun the works; but they were suddenly suspended, and fever and plague * continued to decimate the population. The citizens of London seemed to be as far as ever from their supply of pure water. At this juncture, when all help seemed to fail, and when men were asking each other " who is to do this great work, and

---

* The plague was then a frequent visitor in the city. Numerous proclamations were made by the Lord Mayor and Corporation on the subject,—proclamations ordering wells and pumps to be drawn, and streets to be cleaned,—and precepts for removing hogs out of London, and against the selling or eating of pork. Wherever the plague was in a house, the inhabitant thereof was enjoined to set up outside a pole of the length of seven feet, with a bundle of straw at the top, as a sign that the deadly visitant was within. Wife, children, and servants belonging to that house must carry white rods in their hands for thirty-six days before they were considered "purged." It was also ordered subsequently, that on the street-door of every house infected, or upon a post thereby, the inhabitant must exhibit imprinted on paper a token of St. Anthony's Cross, otherwise called the sign of the Taw **T**, that all persons might have knowledge that such house was infected.—' Corporation of City of London Records,' jor. 12, fol. 136, No. 1. Years 1590 to 1694.

how is it to be done?" citizen Hugh Myddelton, impatient of further delay, came forward and boldly offered to execute it at his own cost. Yet Hugh Myddelton was not an engineer, nor even an architect nor a builder. What he really was, we now proceed briefly to relate, according to the

Myddelton's House at Galch-hill, Denbigh.

best information that we have been able to bring together on the subject.

Hugh Myddelton, the London goldsmith, was born in the year 1555, at Galch-hill, near Denbigh, in North Wales. Richard Myddelton, of Galch-hill, was governor of Denbigh Castle in the reigns of Edward VI., Mary, and Elizabeth. He was a man eminent for his uprightness and integrity,

and is supposed to have been the first member
who sat in Parliament for the town of Denbigh.
His wife was one Jane Dryhurst, the daughter
of an alderman of the town, by whom he had a
family of nine sons and seven daughters.   He
was buried with his wife in the parish church
of Denbigh, called Whitchurch or St. Marcellus ;
where a small monumental brass, placed within

Fac-simile of the Myddelton Brass in Whitchurch Porch.

the porch, represents Richard Myddelton and Jane
his wife, with their sixteen children behind them,
all kneeling.

Several of the Governor's sons rose to distinction.   The third son, William, was one of
Queen Elizabeth's famous sea captains.   He was
also a man of literary tastes, being the author
of a volume entitled ' Barddonnaeth, or the Art of
Welsh Poetry.'   While on his cruises, he occupied

himself in translating the Book of Psalms into Welsh; he finished it in the West Indies, and it was published in 1603, shortly after his death. The fourth son, Thomas, was an eminent citizen and grocer of London. He served the office of Sheriff in 1603, when he was knighted; and he was elected Lord Mayor in 1613. He was the founder of the Chirk Castle family, now represented by Mr. Myddelton Biddulph. The fifth son, Charles, succeeded his father as governor of Denbigh Castle, and when he died bequeathed numerous legacies for charitable uses. The sixth son was Hugh, the subject of this memoir. Robert, the seventh, was, like two of his brothers, a citizen of London, and afterwards a member of Parliament. Foulk, the eighth son, served as high sheriff of the county of Denbigh. This was certainly a large measure of worldly prosperity and fame to fall to the lot of one man's offspring.

Hugh Myddelton was sent up to London to be bred to business there, under the eye of his elder brother Thomas, the grocer and merchant adventurer. In those days country gentlemen of moderate income were accustomed to bind their sons apprentices to merchants, especially where the number of younger sons was large, as it certainly was in the case of Richard Myddelton of Galchhill. There existed at that time in the metropolis numerous exclusive companies or guilds, the admission into which was regarded as a safe road to fortune. The merchants were few in number, constituting almost an aristocracy in themselves; indeed, they were not unfrequently elevated to the peerage because of their wealth as well as public services, and not a few of our present noble families

can trace their pedigree back to some wealthy
skinner, mercer, or tailor, of the reigns of James
or Elizabeth.

Hugh Myddelton was entered an apprentice of
the guild of the Goldsmiths' Company. Having
thus set his son in the way of well-doing, Richard
Myddelton left him to carve out his own career,
relying upon his own energy and ability. He had
done the same with Thomas, whom he had helped
until he could stand by himself; and William,
whom he had educated at Oxford as thoroughly
as his means would afford. These sons having
been fairly launched upon the world, he bequeathed
the residue of his property to his other sons and
daughters.

The goldsmiths of that day were not merely
dealers in plate, but in money. They had succeeded
to much of the business formerly carried on by the
Jews and Venetian merchants established in or near
Lombard-street. They usually united to the trade
of goldsmith that of banker, money-changer, and
money-lender, dealing generally in the precious
metals, and exchanging plate and foreign coin for
gold and silver pieces of English manufacture,
which had become much depreciated by long use
as well as by frequent debasement. It was to the
goldsmiths that persons in want of money then
resorted, as they would now resort to money-
lenders and bankers; and their notes or warrants
of deposit circulated as money, and suggested the
establishment of a bank-note issue, similar to our
present system of bullion and paper currency.
They held the largest proportion of the precious
metals in their possession; hence, when Sir
Thomas Gresham, one of the earliest bankers.

died, it was found that the principal part of his
wealth was comprised in gold chains.*

The place in which Myddelton's goldsmith's
shop was situated was in Bassishaw (now called
Basinghall) Street, and he lived in the overhanging
tenement above it, as was then the custom of city
merchants. Few, if any, lived away from their
places of business. The roads into the country,
close at hand, were impassable in bad weather,
and dangerous at all times. Basing Hall was only
about a bow-shot from the City Wall, beyond which
lay Finsbury Fields, the archery ground of London,
which extended from the open country to the very
wall itself, where stood Moor Gate. The London
of that day consisted almost exclusively of what
is now called The City; and there were few or no
buildings east of Aldgate, north of Cripplegate, or
west of Smithfield. At the accession of James I.
there were only a few rows of thatched cottages in
the Strand, along which, on the river's side, the boats
lay upon the beach. At the same time there were
groves of trees in Finsbury and green pastures in
Holborn; Clerkenwell was a village; St. Pancras
boasted only of a little church standing in meadows;
and St. Martin's, like St. Giles's, was literally "in
the fields." All the country to the west was farm
and pasture land; and woodcocks and partridges
flew over the site of the future Regent Street, May
Fair, and Belgravia.

The population of the city was about 150,000,
living in some 17,000 houses, brick below and
timber above, with picturesque gable-ends, and
sign boards swinging over the footways. The

---

* It may be remembered that Rubens was accustomed to be
paid for his pictures by so many links of gold chain.

upper parts of the houses so overhung the foundations, and the streets were so narrow, that D'Avenant said the opposite neighbours might shake hands without stirring from home. The ways were then quite impassable for carriages, which had not yet indeed been introduced into England; all travelling being on foot or on horseback. When coaches were at length introduced and became fashionable, the aristocracy left the city, through the streets of which their carriages could not pass, and migrated westward to Covent Garden and Westminster.

Those were the days for quiet city gossip and neighbourly chat over matters of local concern; for London had not yet grown so big or so noisy as to extinguish that personal interchange of views on public affairs which continues to characterise most provincial towns. Merchants sat at their doorways in the cool of the summer evenings, under the overhanging gables, and talked over the affairs of trade; whilst those courtiers who still had their residences within the walls, lounged about the fashionable shops to hear the city gossip and talk over the latest news. Myddelton's shop appears to have been one of such fashionable places of resort, and the pleasant tradition was long handed down in the parish of St. Matthew, Friday-street, that Hugh Myddelton and Walter Raleigh used to sit together at the door of the goldsmith's shop, and smoke the newly introduced weed, tobacco, greatly to the amazement of the passers by. It is not improbable that Captain William Myddelton, who lived in London * after his return from the Spanish main

* He resided at the old Elizabethan house in Highgate, afterwards occupied as an inn, called the "King's Head."

in 1591, formed an occasional member of the
group; for Pennant states that he and his friend
Captain Thomas Price, of Plâsgollen, and another,
Captain Koet, were the first who smoked, or as
they then called it, "drank" tobacco publicly in
London, and that the Londoners flocked from all
parts to see them.*

Hugh Myddelton did not confine himself to the
trade of a goldsmith, but from an early period his
enterprising spirit led him to embark in ventures
of trade by sea; and hence, when we find his name
first mentioned in the year 1597, in the records of
his native town of Denbigh, of which he was an
alderman and "capitall burgess," as well as the
representative in Parliament, he is described as
"Cittizen and Gouldsmythe of London, and one
of the Merchant Adventurers of England."† The
trade of London was as yet very small, but a
beginning had been made. A charter was granted
by Henry VII., in 1505, to the Company of Merchant Adventurers of England, conferring on them
special privileges. Previous to that time, almost
the whole trade had been monopolised by the Steelyard Company of Foreign Merchants, whose exclusive privileges were formally withdrawn in 1552.
But for want of an English mercantile navy, the
greater part of the foreign carrying trade of the
country continued long after to be conducted by
foreign ships.

The withdrawal of the privileges of the foreign
merchants in England had, however, an immediate effect in stimulating the home trade, as is
proved by the fact, that in the year following the

* 'Tour in Wales,' vol. ii., p. 31. Ed. 1784.
† Williams's 'Ancient and Modern Denbigh,' p. 105.

suppression of the foreign company, the English Merchant Adventurers shipped off for Flanders no less than 40,000 pieces of cloth. Myddelton entered into this new trade of cloth-making with great energy, and he prosecuted it with so much success, that in a speech delivered by him in the House of Commons on the proposed cloth patent, he stated that he and his partner then maintained several hundred families by that trade. He also seems to have taken part in the maritime adventures of the period, most probably encouraged thereto by his intimacy with Raleigh and other sea captains, including his brother William, who had made profitable ventures on the Spanish main. In short, Hugh Myddelton was regarded as an eminently prosperous man.

At this stage of his affairs, when arrived at a comparatively advanced age, Myddelton took to himself a wife; and the rank and fortune of the lady he married afford some indication of the position he had by this time attained. She was Miss Elizabeth Olmstead, the daughter and sole heiress of John Olmstead of Ingatestone, Essex, with whom the thriving goldsmith and merchant adventurer received a considerable accession of property. That he had secured the regard of his neighbours, and did not disdain to serve them in the local offices to which they chose to elect him, is apparent from the circumstance that he officiated for three years as churchwarden for the parish of St. Matthew, to which post he was appointed in the year 1598.

Myddelton continued to keep up a friendly connection with his native town of Denbigh, and he seems to have been mainly instrumental in obtaining for the borough its charter of incorporation in the reign of Elizabeth. In return for this service the burgesses elected him their first alderman, and in that capacity he signed the first by-laws of the borough in 1597. On the back of the document are some passages in his hand-writing, commencing with "Tafod aur yngenau dedwydd" [A golden tongue is in the mouth of the blessed], followed by other aphorisms, and concluding with some

Whitchurch, or St. Marcellus, Denbigh.

expressions of regret at parting with his brethren, the burgesses of Denbigh, whom he had specially visited on the occasion. It would appear, from subsequent letters of his, that about this time he temporarily resided in the town,—most probably during an attempt which he made to sink for coal in the neighbourhood, which turned out a total failure.

A few years later, Myddelton was appointed Recorder of the borough, and in 1603 he was elected to represent it in Parliament. In those days the office of representative was not so much coveted as it is now, and boroughs remote from the metropolis were occasionally under the necessity of paying their members to induce them to serve. It was, doubtless, an advantage to the burgesses of Denbigh that they had such a man to represent them as Hugh Myddelton, resident in London, and who was moreover an alderman and a benefactor of the town. His two brothers —Thomas Myddelton, citizen and grocer, and Robert, citizen and skinner, of London—were members of the same Parliament, and we find Hugh and Robert frequently associated on committees of inquiry into matters connected with trade and finance. Among the first committees to which the brothers were appointed was one on the subject of a bill for explanation of the Statute of Sewers, and another for the bringing of a fresh stream of running water from the river of Lea, or Uxbridge, to the north parts of the city of London. Thus the providing of a better supply of water to the inhabitants of the metropolis came very early under his notice, and doubtless had some influence in directing his future action on the subject.

At the same time the business in Bassishaw-street was not neglected, for, shortly after the arrival of King James in London, we find Myddelton supplying jewelry for Queen Anne, whose rage for finery of that sort was excessive. A warrant, in the State-Paper-office, orders 250l. to be paid to Hugh Myddelton, goldsmith, for a jewel given by

James I. to the Queen ; * and it is probable that this connection with the Court introduced him to the notice of the King, and facilitated his approach to him when he afterwards had occasion to solicit His Majesty's assistance in bringing the New River works to completion.

The subject of water supply to the northern parts of the city was still under the consideration of parliamentary committees, of which Myddelton was invariably a member ; and at length a bill passed into law, and the necessary powers were conferred. But no steps were taken to carry them into effect. The chief difficulty was not in passing the Act, but in finding the man to execute the work. A proposal made by one Captain Colthurst to bring a running stream from the counties of Hertford and Middlesex, was negatived by the Common Council in 1608. Fever and plague from time to time decimated the population, and the citizens of London seemed as far as ever from being supplied with pure water.

It was at this juncture that Hugh Myddelton stepped forth and declared that if no one else would undertake it, he would, and bring the water from Hertfordshire into London. "The matter,"

---

* " 26th of February, 1604. To Hugh Middleton, Goldsmith, the sum of 250/. for a pendant of one diamond bestowed upon the Queen by His Majesty. By writ dated 9th day of January, 1604, 250/."—Extract from the 'Pell Records.' [The sum named would be equivalent to about 1000/. of our present money. The Queen's passion for jewels may be inferred from the circumstance stated by Dr. Steven in his ' Memoir of George Heriot,' the King's goldsmith (founder of Heriot's Hospital, Edinburgh), that during the ten years which immediately preceded the accession of King James to the throne of Great Britain, Heriot's bills for the Queen's jewels alone could not amount to less than 50,000/. sterling.]

THE NEW RIVER.

quaintly observes Stow, "had been well-mentioned though little minded, long debated but never concluded, till courage and resolution lovingly shook hands together, as it appears, in the soule of this no way to be daunted, well-minded gentleman." When all others held back—lord mayor, corporation, and citizens—Myddelton took courage, and showed what one strong practical man, borne forward by resolute will and purpose, can do. "The dauntless Welshman," says Pennant, "stept forth and smote the rock, and the waters flowed into the thirsting metropolis."

Myddelton's success in life seems to have been attributable not less to his quick intelligence than to his laborious application and indomitable perseverance. He had, it is true, failed in his project of finding coal at Denbigh; but the practical knowledge which he acquired, during his attempt, of the arts of mining and excavation, had disciplined his skill and given him fertility of resources, as well as cultivated in him that power of grappling with difficulties, which emboldened him to undertake this great work, more like that of a Roman emperor than of a private London citizen.

The corporation were only too glad to transfer to him the powers with which they had been invested by the legislature, together with the labour, the anxiety, the expense, and the risk of carrying out an undertaking which they regarded as so gigantic. On the 28th of March, 1609, the corporation accordingly formally agreed to his proposal to bring a supply of water from Amwell and Chadwell, in Hertfordshire, to Islington, as being "a thing of great consequence, worthy of acceptation for the good of the city;" but subject to his beginning the

works within two months from the date of their
acceptance of his offer, and doing his best to finish
the same within four years. A regular indenture
was drawn up and executed between the parties on
the 21st of April following; and Myddelton began
the works and "turned the first sod" in the course
of the following month, according to the agreement. The principal spring was at Chadwell, near
Ware, and the operations commenced at that point.
The second spring was at Amwell, near the same
town; each being about twenty miles from London
as the crow flies.

The general plan adopted by Myddelton in
cutting the New River was to follow a contour
line, as far as practicable, from the then level
of the Chadwell Spring to the circular pond at
Islington, subsequently called the New River
Head. The stream originally presented a fall of
about 2 inches in the mile, and its City end was at
the level of about 82 feet above what is now known
as Trinity high water mark. Where the fall of the
ground was found inconveniently rapid, a stop-gate
was introduced across the stream, penning from
3 to 4 feet perpendicularly, the water flowing over
weirs down to the next level.

To accommodate the cut to the level of the
ground as much as possible, numerous deviations
 were made, and
the river was led
along the sides
of the hills, from
which sufficient
soil was exca-
vated to form the lower bank of the intended
stream. Each valley was traversed on one side until

it reached a point where it could be crossed; and there an embankment became necessary, in some cases of from 8 to 10 feet in height, along the top of which the water was conducted in a channel of the proper dimensions. In those places where the embankments were formed, provision had of course to be made for the passage of the surface waters from the west of the line of works into the river

The Boarded River formerly at Bush Hill.

Lea, which forms the natural drain of the district. In some cases the drainage waters were conveyed under the New River in culverts, and in others over it by what were termed flashes. At each of the "flashes" there were extensive swamps, where the flood-waters were upheld to such a level as to enable them to pass over the flash, which consisted of a wooden trough, about twelve feet wide and three deep, extending across the river; and from these swamps, as well as from every other running stream, such apparatus was introduced as enabled the Company to avail themselves of the supply of water which they afforded, when required. Openings were also left in the banks for the passage of roads under the stream, the continuity of which was in such cases maintained either by arches or timber troughs lined with lead. One of these

Brick Arch under the New River, formerly near Bush Hill.

troughs, at Bush Hill, near Edmonton, was about 660 feet long, and 5 feet deep. A brick arch also formed part of this aqueduct, under which flowed a stream which had its source in Enfield Chase; the arch sustaining the trough and the

---

The map on the next page will enable the reader to trace the line of the New River works between Amwell, Chadwell, and London. The dotted lines indicate those parts of the old course which have since been superseded by more direct cuts, represented by the continuous black line. Where the loops have been detached from the present line of works, they are, in most instances, laid dry, and may be examined and measured correctly, as also the soil of which the banks were originally formed.

Map of the New River.

road along its side. Another strong timber aqueduct, 460 feet long and 17 feet high, conducted the New River over the valley near where it entered the parish of Islington. This was long known in the neighbourhood as "Myddelton's Boarded River." At Islington also there was a brick tunnel of considerable extent, and another at Newington. That at Islington averaged in section about 3 feet by 5, and appears to have been executed at different periods, in short lengths. Such were the principal works along the New River. Its original extent was much greater than it is at present, from its frequent windings along the high grounds for the purpose of avoiding heavy cuttings and embankments. Although the distance between London and Ware is only about 20 miles, the New River, as originally constructed, was not less than 38¾ miles in length.

The works were no sooner begun than a swarm of opponents sprang up. The owners and occupiers of lands through which the New River was to be cut, strongly objected to it as most injurious to their interests. In a petition presented by them to Parliament, they alleged that their meadows would be turned into "bogs and quagmires," and arable land become "squallid ground;" that their farms would be "mangled" and their fields cut up into quillets and "small peeces;" that the "cut," which was no better than a deep ditch, dangerous to men and cattle, would, upon "soden raines," inundate the adjoining meadows and pastures, to the utter ruin of many poor men; that the church would be wronged in its tithe without remedy; that the highway between London and Ware would be made impassable; and that an infinity of evils

would be perpetrated and irretrievable injuries
inflicted on themselves and their posterity. The
opponents also pointed out that the Mayor and
corporation would have nothing to do with the
business, but, by an irrevocable act of the Common
Council, had transferred their powers of executing
the works to Mr. Myddelton and his heirs, "who
doth the same for his own private benefit."

The agitation against the measure was next
taken up in Parliament. "Much ado there is in
the House," writes Mr. Beaulieu, on the 9th of
May, 1610, to a friend in the country, "about the
work undertaken, and far advanced already by
Myddelton, of the cutting of a river and bringing
it to London from ten or twelve miles off, through
the grounds of many men, who, for their particular
interests, so strongly oppose themselves to it, and
are like (as it is said) to overthrow it all." On the
20th of June following, a Bill was introduced and
committed to repeal the Act authorising the construction of the New River. A committee of ten
was appointed a few days after " to view " the river
and to certify respecting the progress made with
the works, doubtless with the object of ascertain-
ing what damage had actually been done, or was
likely to be done, to private property. The committee were directed to make their report in the
next session ; but as Parliament was prorogued in
July, and did not meet for four years, the subject
is not again mentioned in the Journals of the
House.

Worse than all, was the popular opposition
which Myddelton had to encounter. The pastor
of Tottenham, writing in 1631, speaks of the New
River as " brought with an ill wille from Ware to

London." Stow, who was a contemporary and enthusiastic admirer of Myddelton, says bitterly, "If those enemies of all good endeavours, Danger, Difficulty, Impossibility, Detraction, Contempt, Scorn, Derision, yea, and Desperate Despight, could have prevailed, by their accursed and malevolent interposition, either before, at the beginning, in the very birth of the proceeding, or in the least stolne advantage of the whole prosecution, this Worke, of so great worth, had never been accomplished." Stow records that he rode down divers times to see the progress made in cutting and constructing the New River, and "diligently observed that admirable art, pains, and industry were bestowed for the passage of it, by reason that all grounds are not of a like nature, some being oozy and very muddy, others again as stiff, craggy, and stony. The depth of the trench," he adds, "in some places descended full thirty feet, if not more, whereas in other places it required a sprightful art again to mount it over a valley in a trough, between a couple of hills, and the trough all the while borne up by wooden arches, some of them fixed in the ground very deep, and rising in height above twenty-three feet."

It shortly became apparent to Myddelton that the time originally fixed by the Common Council for the completion of the works had been too short, and we accordingly find him petitioning the Corporation for its extension. This was granted him for five years more, on the ground of the opposition and difficulties which had been thrown in his way by the occupiers and landowners along the line of the proposed stream. It has usually been alleged that Myddelton fell short of funds, and that the Corporation refused him the necessary pecuniary assistance ; but the Corporation records do not bear out this statement, the only application apparently made by Myddelton being for an extension of time. It has also been stated that he was opposed by the water-carriers, and that they even stirred up the Corporation to oppose the construction of the New River; but this statement seems to be equally without foundation. The principal obstacle which Myddelton had to encounter was unquestionably the opposition of the landowners and occupiers ; and it was so obstinate that in his emergency he was driven to apply to the King for assistance.

Though James I. may have been ridiculous and unkingly in many respects, he nevertheless appears throughout his reign to have exhibited a sensible desire to encourage the industry and develop the resources of the kingdom he governed. It was he who made the right royal declaration with reference to the drowned lands in the Fens, that he would not suffer the waters to retain their dominion over the soil which skill and labour might reclaim for human uses. He projected the drainage and reclamation of the royal manor of Hatfield Chase, as well as the reclamation of Sedgemoor and Malvern Chase ; and when the landowners in the Fens would take no steps to drain the Great Level, he expressed the determination to become himself the sole undertaker. And now, when Hugh Myddelton's admirable project for supplying the citizens of London with water threatened to break down by reason of the strong local opposition offered to it, and while it was spoken of by many with derision and contempt as an impracticable undertaking, the same monarch came to his help, and while he rescued Myddelton from heavy loss, it might be ruin, he enabled him to prosecute his important enterprise to completion.

James had probably become interested in the works from observing their progress at the point at which they passed through the Royal Park at Theobalds, a little beyond Enfield.* Theobalds was the favourite residence of the King, where he frequently indulged in the pastime of hunting ; and on passing the labourers occupied in cutting the New River, he would naturally make inquiries as to their progress. The undertaking was of a character so unusual, and so much of it passed directly through the King's domains, that he could not but be curious about it. Myddelton, having had dealings with His Majesty as a jeweller, seized the opportunity of making known his need of immediate help, otherwise the project must fall through. Several interviews took place between them at Theobalds and on the ground ; and the result was that James determined to support the engineer with his effective help as King, and also with the help of the State purse, to enable the work to be carried out.

An agreement was accordingly entered into between the King and Myddelton, the original of which is deposited in the Rolls-office, and is a

---

* Theobalds, a singularly beautiful place, where Elizabeth held counsel with Burleigh, James often lived, and Charles played with his children. The palace was ordered to be pulled down by the Long Parliament, in spite of the commissioners' report that it was "an excellent building in very good repair ; " and, the materials having been sold to the highest bidder, the proceeds were divided amongst the soldiers of Cromwell and Fairfax. The materials alone realised not less than 8275*l.* 11*s.*

highly interesting document. It is contained on
seven skins, and is very lengthy ; but the following
abstract will sufficiently show the nature of the
arrangement between the parties. The Grant, as
it is described, is under the Great Seal, and dated
the 2nd of May, 1612. It is based upon certain
articles of agreement, made between King James I.
and Hugh Myddelton, "citizen and goldsmith of
London," on the 5th of November preceding. It
stipulates that His Majesty shall discharge a
moiety of all necessary expenses for bringing the
stream of water within "one mile of the city," as
well as a moiety of the disbursements "already
made" by Hugh Myddelton, upon the latter sur-
rendering an account, and swearing to the truth
of the same. In consideration of His Majesty's
pecuniary assistance, Myddelton assigned to him
a moiety of the interest in, and profits to arise
from, the New River " for ever," with the exception of a small quill or pipe of water which the
said Myddelton had granted, at the time of his
agreement with the City, to the poor people inhabiting St. John-street and Aldersgate-street,—
which exception His Majesty allowed.

One of the first benefits Myddelton derived
from the arrangement was the repayment to him
of one-half the expenditure which had been incurred to that time. It appears from the first
certificate delivered to the Lord Treasurer, that
the total expenditure to the end of the year 1612
had been 4485l. 18s. 11d., as attested by Hugh
Myddelton, acting on his own behalf, and Miles
Whitacres acting on behalf of the King. Further
payments were made out of the Treasury for costs
disbursed in executing the works ; and it would

appear from the public records that the total payments made out of the Royal Treasury on account
of the New River works amounted to 8609*l*. 14*s*. 6*d*.
As the books of the New River Company were
accidentally destroyed by a fire many years ago,
we are unable to test the accuracy of these figures
by comparison with the financial records of the
Company ; but, taken in conjunction with other
circumstances hereafter to be mentioned, the amount
stated represents, with as near an approach to
accuracy as can now be reached, one-half of the
original cost of constructing the New River.

As the undertaking proceeded, with the powerful
help of the King and the public Treasury, and
as the great public uses of the New River began
to be recognised, the voice of derision became
gradually stilled, and congratulations began to
rise up on all sides in view of the approaching
completion of the bold enterprise. The scheme
had ceased to be visionary, as it had at first
appeared, for the water was already brought
within a mile of Islington ; all that was wanted
to admit it to the reservoir being the completion
of the tunnel near that place. At length that
too was finished ; and now King, Corporation,
and citizens vied with each other in doing honour
to the enterprising and public spirited Hugh
Myddelton. The Corporation elected his brother
Thomas Lord Mayor for the year ; and on Michael-
mas Day, 1613, the citizens assembled in great
numbers to celebrate by a public pageant the admission of the New River water to the metropolis.
The ceremony took place at the new cistern at
Islington, in the presence of the Lord Mayor,
Aldermen, Common Council, and a great concourse

of spectators. A troop of some three score labourers in green Monmouth caps, bearing spades and mattocks, or such other implements as they had used in the construction of the work, marched round the cistern to the martial music of drums and trumpets, after which a metrical speech, composed by one Thomas Middleton, was read aloud, expressive of the sentiments of the workmen. The following extract may be given, as showing the character of the persons employed on the undertaking :—

> First, here's the Overseer, this try'd man,
> An antient souldier and an artizan ;
> The Clearke ; next him the Mathematian ;
> The Maister of the Timber-worke takes place
> Next after these ; the Measurer in like case ;
> Bricklayer, and Enginer ; and after those
> The Borer, and the Pavier; then it showes
> The Labourers next ; Keeper of Amwell Head ;
> The Walkers last ;—so all their names are read.
> Yet these but parcels of six hundred more,
> That, at one time, have been imploy'd before ;
> Yet these in sight and all the rest will say
> That all the weeke they had their Royall pay !

At the conclusion of the recitation the flood-gates were thrown open, and the stream of pure water rushed into the cistern amidst loud huzzas, the firing of mortars, the pealing of bells, and the triumphant welcome of drums and trumpets.*

* A large print was afterwards published by G. Bickham, in commemoration of the event, entitled 'Sir Hugh Myddelton's Glory.' It represents the scene of the ceremony, the reservoir, with the stream rushing into it ; the Lord Mayor (Sir John Swinnerton) on a white palfrey, pointing exultingly to Sir Hugh ; the Recorder, Sir Henry Montague, afterwards Lord Keeper and Earl of Manchester, and by his side the Lord Mayor elect, the projector's brother, Maister Thomas Myddelton. Various figures gesticulating their admiration occupy the foreground, whilst the

It is rather curious that James I. was afterwards nearly drowned in the New River which he had enabled Hugh Myddelton to complete. He had gone out one winter's day after dinner to ride in the park at Theobalds accompanied by his son Prince Charles; when, about three miles from the palace, his horse stumbled and fell, and the King was thrown into the river. It was slightly frozen over at the time, and the King's body disappeared under the ice, nothing but his boots remaining visible. Sir Richard Young rushed in to his rescue, and dragged him out, when "there came much water out of his mouth and body." He was, however, able to ride back to Theobalds, where he got to bed and was soon well again. The King attributed his accident to the neglect of Sir Hugh and the Corporation of London in not taking measures to properly fence the river, and he did not readily forget it; for when the Lord Mayor, Sir Edward Barkham, accompanied by the Recorder, Sir Heneage Finch, attended the King at Greenwich, in June, 1622, to be knighted, James took occasion, in rather strong terms, to remind the Lord Mayor and his brethren of his recent mischance in "Myddelton's Water."

It is scarcely necessary to point out the great benefits conferred upon the inhabitants of London by the construction of the New River, which furnished them with an abundant and unremitting supply of pure water for domestic and other purposes. Along this new channel were poured into the city several millions of gallons daily; and

foot of the print is garnished with little "chambers," or miniature mortars, spontaneously exploding. There is a copy of the original print in the British Museum.

the reservoirs at New River Head being, as before
stated, at an elevation of 82 feet above the level of
high water in the Thames, they were thus capable of
supplying through pipes the basement stories of the
greater number of houses then in the metropolis.

The pipes which were laid down in the first
instance to convey the water to the inhabitants
were made of *wood*, prin-
cipally elm ; and at one
time the New River
Company had wooden
pipes laid down through
the streets to the ex-
tent of about 400 miles!
But the leakage was so
great through the por-
ousness of the material,
—about one-fourth of the
whole quantity of water
supplied passing away
by filtration,—and the
decay of the pipes in
ordinary weather was so

" New River Water !"
[After an Ancient Print.]

rapid, besides being liable to burst during frosts,
that they were ultimately abandoned when mechanical skill was sufficiently advanced to enable
pipes of cast-iron to be substituted for them. For
a long time, however, a strong prejudice existed
against the use of water conveyed through pipes
of any kind, and the cry of the water carriers
long continued to be familiar to London ears, of
"Any New River water here!" "Fresh and fair
New River water! none of your pipe sludge!"

Among the many important uses to which the
plentiful supply of New River water was put, was

the extinction of fires, then both frequent and destructive, in consequence of the greater part of the
old houses in London being built of wood. Stow
particularly mentions the case of a fire which broke
out in Broad Street, on the 12th November, 1623,
in the house of Sir William Cockaigne, which
speedily extended itself to several of the adjoining buildings.   We are told by the chronicler,
that "Sir Hugh Myddelton, upon the first knowledge thereof, caused all the sluices of the water-
cisterne in the field to be left open, whereby there
was plenty of water to quench the fire. The water"
[of the New River], he continues, "hath done many
like benefits in sundrie like former distresses."

We now proceed to follow the fortunes of Myddelton in connexion with the New River Company.
The year after the public opening of the cistern at
Islington, we find him a petitioner to the Corporation for a loan of 3000*l.*, for three years, at six per
cent., which was granted him "in consideration of
the benefit likely to accrue to the city from his
New River;" his sureties being the Lord Mayor
(Hayes), Mr. Robert Myddelton (his brother), and
Mr. Robert Bateman.   There is every reason to
believe that Myddelton had involved himself in
difficulties by locking up his capital in this costly
undertaking ; and that he was driven to solicit the
loan to carry him through until he had been enabled
to dispose of the greater part of his interest in
the concern to other capitalists.   This he seems to
have done very shortly after the completion of the
works.   The capital was divided into seventy-two
shares,* one-half of which belonged to Myddelton

---

* In Pennant's 'London' it is stated that the original shares in
the concern were 100*l.* each, whereas Entick makes them to have

ANOTHER VIEW OF THE NEW RIVER.

[*To face p. 96.*

and the other half to the King, in consideration of the latter having borne one-half of the cost.   Of the thirty-six shares owned by the former, as many as twenty-eight were conveyed by him to other persons; and that he realised a considerable sum by the sale is countenanced by the circumstance that we find him shortly after embarked in an undertaking hereafter to be described, requiring the command of a very large capital.

The shareholders were incorporated by letters patent on the 21st of June, 1619, under the title of "The Governors and Company of the New River brought from Chadwell and Amwell to London." *
The government of the corporation was vested in the twenty-nine adventurers who held amongst them the thirty-six shares originally belonging to Myddelton, who had by that time reduced his holding to only two shares.   At the first Court of proprietors, held on the 2nd of November, 1619, he was appointed Governor, and Robert Bateman Deputy-Governor of the Company.   Sir Giles Mompesson was appointed, on behalf of the King, Surveyor of the profits of the New River, with authority to attend the meetings, inspect the accounts, &c., with a grant for such service of 200*l.* per annum out of the King's moiety of the profits of the said river. It was long, however, before there were any profits

amounted to not less than 7000*l.* each !   This is only another illustration of the hap-hazard statements put forth respecting Sir Hugh and his works.   The original cost of the New River probably did not amount to more than 18,000*l.*, in which case the capital represented by each original share would be 250*l.*

* On the occasion of its subsequent confirmation by parliament, Sir Edward Coke said : "This is a very good bill, and prevents one great mischief that hangs over the city.   *Nimis potatio : frequens incendium.*"

to be divided; for the cost of making repairs and
improvements, and laying down wooden pipes,
continued to be very great for many years; and
the ingenious method of paying dividends out of
capital, to keep up the price of shares and invite
further speculation, had not yet been invented. In
fact, no dividend whatever was paid until after the
lapse of twenty years from the date of opening
the New River at Islington; and the first dividend
only amounted to 15*l*. 3*s*. 3*d*. a share. The next
dividend of 3*l*. 4*s*. 2*d*. was paid three years later,
in 1636; and as the concern seemed to offer no
great prospect of improvement, and a further
call on the proprietors was expected, Charles I.,
who required all his available means for other
purposes, finally regranted his thirty-six " King's
shares" to the Company, under his great seal, in
consideration of a fee farm rent of 500*l*., which
is to this day paid by them yearly into the King's
exchequer.

Notwithstanding this untoward commencement
of the New River Company, it made great and
rapid progress when its early commercial difficulties had been overcome; and after the year 1640
its prosperity steadily kept pace with the population and wealth of the metropolis. By the end
of the seventeenth century the dividend paid was
at the rate of about 200*l*. per share; at the end
of the eighteenth century the dividend was above
500*l*. per share; and at the present date each share
produces about 850*l*. a year. At only twenty
years' purchase, the capital value of a single share
at this day would be about 17,000*l*. But most of
the shares have in course of time, by alienation and
bequeathment, become very much subdivided; the

possessors of two or more fractional parts of a share being enabled, under a decree of Lord Chancellor Cowper, in 1711, to depute a person to represent them in the government of the Company.

Source of the New River at Chadwell, near Ware, with the Monumental Pedestal in memory of Sir Hugh Myddelton.

CHAPTER IV.

HUGH MYDDELTON (*continued*)—HIS OTHER ENGINEER-
ING AND MINING WORKS—AND DEATH.

SHORTLY after the completion of the New River,
and the organisation of the Company for the supply
of water to the metropolis, we find Hugh Myddelton
entering upon a new and formidable enterprise—
that of enclosing a large tract of drowned land from
the sea.

The scene of his operations on this occasion
was the eastern extremity of the Isle of Wight,
at a place now marked on the maps as Brading
Harbour. This harbour or haven consists of a
tract of about eight hundred acres in extent. At
low water it appears a wide mud flat, through the
middle of which a small stream, called the Yar,
winds its way from near the village of Brading, at
the head of the haven, to the sea at its eastern
extremity; whilst at high tide it forms a beautiful
and apparently inland lake, embayed between
hills of moderate elevation covered with trees, in
many places down to the water's edge. At its
seaward margin Bembridge Point stretches out
as if to meet the promontory on the opposite
shore, where stands the old tower of St. Helen's
Church, now used as a sea-mark; and, as seen

from most points, the bay seems to be completely landlocked.

View of Brading Haven, temporarily reclaimed by Sir Hugh Myddelton, as seen from the Village of Brading.

The reclamation of so large a tract of land, apparently so conveniently situated for the purpose, had long been matter of speculation.   It is not improbable that at some early period neither swamp nor lake existed at Brading Haven, but a green and fertile valley; for in the course of the works undertaken by Sir Hugh Myddelton for its recovery from the sea, a well, strongly cased with stone, was discovered near the middle of the haven, indicating the existence of a population formerly settled on the soil.   The sea must, however, have burst in and destroyed the settlement, laying the whole area under water.

In King James's reign, when the inning of drowned lands began to receive an unusual degree of attention, the project of reclaiming Brading Haven was again revived; and in the year 1616 a grant was made of the drowned district to one John Gibb, the King reserving to himself a rental of 20*l*. per annum.   The owners of the adjoining lands contested the grant, claiming a prior right to the property in the haven, whatever its worth might be.   But the verdict of the Exchequer went against

Map of Brading Harbour.

the landowners, and the right of the King to grant
the area of the haven for the purpose of reclamation was maintained. It appears that Gibb sold
his grant to one Sir Bevis Thelwall, a page of the
King's bedchamber, who at once invited Hugh
Myddelton to join him in undertaking the work;
but Thelwall would not agree to pay Gibb anything until the enterprise had been found practicable. In 1620 we find that a correspondence was
in progress as to "the composition to be made
by the Solicitor-General with Myddelton touching the draining of certain lands in the Isle of
Wight, and the bargain having been made according to such directions as His Majesty hath given,
then to prepare the surrender, and thereupon
such other assurance for His Majesty as shall be
requisite." *

A satisfactory arrangement having been made
with the King, Myddelton began the work of reclaiming the haven in the course of the same year.
He sent to Holland for Dutch workmen familiar
with such undertakings; and from the manner in
which he carried out his embankment, it is obvious
that he mainly followed the Dutch method of reclamation, which, as we have already seen in the
case of the drainage of the Fens by Vermuyden,
was not, in many respects, well adapted for English
practice. But it would also appear, from a patent
for draining land which he took out in 1621, that he
employed some invention of his own for the purpose of facilitating the work. The introduction
to the grant of the patent runs as follows :—

---

* 'Domestic Calendar of State Papers.' Docquet, 13th August,
1620.

"WHEREAS wee are given to vnderstand that our welbeloved subiect Hugh Middleton, Citizen and Goldsmith of London, hath to his very great charge maynteyned many strangers and others, and bestowed much of his tyme to invent a new way, and by his industrie, greate charge, paynes, and long experience, hath devised and found out 'A NEW INVENÇON, SKILL, OR WAY FOR THE WYNNING AND DRAYNING OF MANY GROUNDℭ WHICH ARE DAYLIE AND DESPERATELIE SURROUNDED WITHIN OUR KINGDOME OF ENGLAND AND DOMINION OF WALES,' and is now in very great hope to bringe the same to good effect, the same not being heretofore knowne, experimented, or vsed within our said realme or dominion, whereby much benefitt, which as yet is lost, will certenly be brought both to vs in particular and to our comon wealth in generall, and hath offered to publish and practise his skill amongest our loving subiectℭ . . ., KNOWE YEE, that wee, tendring the weale of this our kingdom and the benefitt of our subiectℭ, and out of our princely care to nourish all artℭ, invencions, and studdies whereof there may be any necessary or pffitable vse within our dominions, and out of our desire to cherish and encourage the industries and paynes of all other our loving subiectℭ in the like laudable indeavors, and to recompence the labors and expences of the said Hugh Middleton disbursed and to be susteyned as aforesaid, and for the good opinion wee have conceived of the said Hugh Middleton, for that worthy worke of his in bringing the New River to our cittie of London, and his care and industrie in busines of like nature tending to the publicke good . . . doe give and graunt full, free, and absolute licence, libertie, power, and authoritie vnto the said Hughe Middleton, his deputies," &c. to use and practise the same during the terme of fowerteene years next ensuing the date hereof.

No description is given of the particular method adopted by Myddelton in forming his embankments. It would, however, appear that he proceeded by driving piles into the bottom of the Haven near

Bembridge Point where it is about the narrowest, and thus formed a strong embankment at its junction with the sea, but unfortunately without

Entrance to Brading Harbour, from St. Helen's Old Tower.*

* The above view represents the present state of the entrance to Brading Haven. A wide ridge of drifted sand lies across it, in front of the old bank raised by Myddelton, which extended from a point below the hill under "Mrs. Grant's house," a little to the westward of the village of Bembridge (seen on the opposite shore) to what are now called "The Boat Houses," situated towards the northern side of the haven, and behind the sand-ridge extending across the view. The black piles driven into the bottom of the haven in the process of embankment are still to be seen sticking up at low water; and only a few years since the old gates which served for a sluice were dug up near the Boat Houses. At the extremity of the sand-ridge there is a ferry across to the village of Bembridge, in front of which is the narrow entrance into the haven. There have been serious encroachments of the sea on that side of late years, and the channel has become much impeded; so much so that it has been feared that the navigation would be lost. The old church-tower of St. Helen's, faced with brick and whitewashed, on the right of the view, is still used as a sea-mark.

making adequate provision for the egress of the inland waters.

A curious contemporary manuscript by Sir John Oglander is still extant, preserved amongst the archives of the Oglander family, who have held the adjoining lands from a period antecedent to the date of the Conquest, which we cannot do better than quote, as giving the most authentic account extant of the circumstances connected with the enclosing of Brading Haven by Hugh Myddelton. This manuscript says :—

" Brading Haven was begged first of all of King James by one Mr. John Gibb, being a groom of his bedchamber, and the man that King James trusted to carry the reprieve to Winchester for my Lord George Cobham and Sir Walter Rawleigh, when some of them were on the scaffold to be executed. This man was put on to beg it of King James by one Sir Bevis Thelwall, who was then one of the pages of the bedchamber. After he had begged it, Sir Bevis would give him nothing for it until the haven were cleared; for the gentlemen of the island whose lands join to the haven challenged it as belonging unto them. King James was wonderful earnest in the business, both because it concerned his old servant, and also because it would be a leading case for the fens in Lincolnshire. After the verdict went in the Chequer against the gentlemen, then Sir Bevis Thelwall would give nothing for it till he could see that it was feasible to be inned from the sea ; whereupon one Sir Hugh Myddelton was called in to assist and undertake the work, and Dutchmen were brought out of the Low Countries, and they began to inn the haven about the 20th of December, 1620. Then, when it was taken in, King James compelled Thelwall and Myddelton to give John Gibb (who the King called ' Father ') 2000*l.* Afterwards Sir Hugh Myddelton, like a crafty fox and subtle citizen, put it off wholly to Sir Bevis Thelwall, betwixt whom afterwards there was a great suit in the Chancery; but Sir Bevis did enjoy it

some eight years, and bestowed much money in building of
a barnhouse, mill, fencing of it, and in many other necessary
works.

"But now let me tell you somewhat of Sir Bevis Thelwall
and Sir Hugh Myddelton, and of the nature of the ground
after it was inned, and the cause of the last breach. Sir Bevis
was a gentleman's son in Wales, bound apprentice to a mercer
in Cheapside, and afterwards executed that trade till King
James came into England : then he gave up, and purchased to
be one of the pages of the bedchamber, where, being an understanding man, and knowing how to handle the Scots, did in
that infancy gain a fair estate by getting the Scots to beg for
themselves that which he first found out for them, and then
himself buying of them with ready money under half the value.
He was a very bold fellow, and one that King James very well
affected. Sir Hugh Myddelton was a goldsmith in London.
This and other famous works brought him into the world,
viz., his London waterwork, Brading Haven, and his mine in
Wales.

"The nature of the ground, after it was inned, was not
answerable to what was expected, for almost the moiety of it
next to the sea was a light running sand, and of little worth.
The best of it was down at the farther end next to Brading,
my Marsh, and Knight's Tenement, in Bembridge. I account
that there was 200 acres that might be worth 6s. 8d. the acre,
and all the rest 2s. 6d. the acre. The total of the haven was
706 acres. Sir Hugh Myddelton, before he sold, tried all
experiments in it : he sowed wheat, barley, oats, cabbage seed,
and last of all rape seed, which proved best ; but all the others
came to nothing. The only inconvenience was in it that the
sea brought in so much sand and ooze and seaweed that choked
up the passage of the water to go out, insomuch as I am of
opinion that if the sea had not broke in Sir Bevis could hardly
have kept it, for there would have been no current for the
water to go out ; for the eastern tide brought so much sand as
the water was not of force to drive it away, so that in time it
would have laid to the sea, or else the sea would have drowned

the whole country. Therefore, in my opinion, it is not good meddling with a haven so near the main ocean.

"The country (I mean the common people) was very much against the inning of it, as out of their slender capacity thinking by a little fishing and fowling there would accrue more benefit than by pasturage; but this I am sure of, it caused, after the first three years, a great deal of more health in these parts than was ever before; and another thing is remarkable, that whereas we thought it would have improved our marshes, certainly they were the worse for it, and rotted sheep which before fatted there.

"The cause of the last breach was by reason of a wet time when the haven was full of water, and then a high spring tide, when both the waters met underneath in the loose sand. On the 8th of March, 1630, one Andrew Ripley that was put in earnest to look to Brading Haven by Sir Bevis Thelwall, came in post to my house in Newport to inform me that the sea had made a breach in the said haven near the easternmost end. I demanded of him what the charge might be to stop it out; he told me he thought 40s., whereupon I bid him go thither and get workmen against the next day morning, and some carts, and I would pay them their wages; but the sea the next day came so forcibly in that there was no meddling of it, for Ripley went up presently to London to Sir Bevis Thelwall himself, to have him come down and take some further course; but within four days after the sea had won so much on the haven, and made the breach so wide and deep, that on the 15th of March when I came thither to see it I knew not well what to judge of it, for whereas at the first 5l. would have stopped it out, now I think 200l. will not do it, and what will be the event of it time will tell. Sir Bevis on news of this breach came into the island on the 17th of March, 1630, and brought with him a letter from my Lord Conway to me and Sir Edward Dennies, desiring us to cause my Lady Worsley, on behalf of her son, to make up the breach which happened in her ground through their neglect. She returned us an answer that she thought that the law would not compel her unto it, and therefore desired to be excused, which answer we returned to my lord. What the event will be I know not, but it seemeth to me not reasonable that she should suffer for not complying with his request. If he had not inned the haven this accident could never have happened; therefore he giving the cause, that she should apply the cure I understand not. But this I am sure, that Sir Bevis thinketh to recover of her and her son all his charges, which he now sweareth every way to be 2000l. For my part, I would wish no friend of mine to have any hand in the second inning of it. Truly all the better sort of the island were very sorry for Sir Bevis Thelwall, and the commoner sort were as glad as to say truly of Sir Bevis that he did the country many good offices, and was ready at all times to do his best for the public and for everyone.

" Sir Hugh Myddelton took it first in, and it was proper for none but him, because he had a mine of silver in Wales to maintain it. It cost at the first taking of it in 4000l., then they gave 2000l. to Mr. John Gibb for it, who had begged it of King James; afterwards, in building the barn and dwelling-house, and water-mill, with the ditching and quick-setting, and making all the partitions, it could not have cost less than 200l. more : so in the total it stood them, from the time they began to take it in, until the 8th of March, a loss of 7000l."

It will thus be observed that the loss of this undertaking fell upon Thelwall, and not upon Myddelton, who sold out of the adventure long before the sea burst through the embankment. The date of conveyance of his rights in the reclaimed land to Sir Bevis Thelwall was the 4th September, 1624, nearly six years before the final ruin of the work. He had, therefore, got his capital out of the concern, most probably with his profit as contractor, and was thus free to embark in the important mining enterprise in Wales, on which we find him next engaged.

Sir Hugh continued to maintain his Parliamentary connection with his native town of Denbigh, of which he was still the representative. We do not find that he took an active part in political questions. The name of his brother, Sir Thomas, frequently appears in the Parliamentary debates of the time, and he was throughout a

Chart of Mines in North Wales.

[From an old Print in the British Museum.]

strong opponent of the Court party; but that of Sir Hugh only occurs in connection with commercial topics or schemes of internal improvement, on which he seems to have been consulted as an authority.

Sir Hugh's occasional visits to his constituents brought him into contact with Welsh families, and

made him acquainted with the mining enterprises
then on foot in different parts of Wales—so rich
in ores of copper, lead, and iron. It appears that
the Governor and Company of Mines Royal in
Cardiganshire were incorporated in the year 1604,
for the purpose of working the lead and silver
mines of that county. The principal were those
at Cwmsymlog and the Darren Hills, situated
about midway, as the crow flies, between Aberystwith and the mountain of Plinlimmon, and at
Tallybout, about midway between Aberystwith and
the estuary at the mouth of the River Dovey.
They were all situated in the township of Skibery
Coed, in the northern part of the county of
Cardigan. For many years these mines (which
were first opened out by the Romans) were
worked by the Corporation of Mines Royal; but
it does not appear that much success attended
their operations. Mining was little understood
then, and all kinds of pumping and lifting machinery
were clumsy and inefficient. Although there was
no want of ore, the mines were so drowned by
water, that the metal could not well be got at and
worked out.

Myddelton's spirit of enterprise was excited by
the prospect of battling with the water and getting
at the rich ore, and he had confidence that his
mechanical ability would enable him to overcome
the difficulties. The Company of Mines Royal
were only too glad to get rid of their unprofitable
undertaking, and they agreed to farm their mines
to Sir Hugh at the rental of 400*l.* per annum.
This was in the year 1617, some time after he
had completed his New River works, but before
he had begun the embankment of Brading Haven,

—and Sir Bevis Thelwall was also a partner with him in this new venture. It took him some time to clear the mines of water, which he did by pumping-machines of his own contrivance; but at length sufficient ore was raised for testing, and it was found to contain a satisfactory proportion of silver. His mining adventure seems to have been attended with success, for we shortly afterwards find him sending considerable quantities of silver to the Royal Mint to be coined.

King James was so much gratified by the further proofs of Myddelton's skill and enterprise, displayed in his embankment of Brading Harbour and his successful mining operations in Wales, that he raised him to the dignity of a Baronet on the 19th of October, 1622; and the compliment was all the more marked by His Majesty directing that Sir Hugh should be discharged from the payment of the customary fees, amounting to 1095*l.*, and that the dignity should be conferred upon him without any charge whatever.* The patent of baronetcy granted on the occasion set forth the "reasons and considerations" which induced the King to confer the honour; and it may not be out of place to remark, that though more eminent industrial services have been

---

* Sloane MS. (British Museum), vol. ii. 4177. Also 'Calendar of Domestic State Papers,' Oct. 19th, 1622: "Grant to Hugh Myddelton of the rank of Baronet, granting discharge of 1095*l.* due on being made a Baronet." The usual statement is to the effect that Myddelton was knighted on the occasion of the opening of the New River in 1613. But this was not the case, as it will be found from the patent for draining land taken out by him in 1621 (see *ante*, p. 103), that he was then described as simply "Citizen and Goldsmith, of London." Nor is his name to be found in any contemporary list of King James's Knights.

THE OAK ROOM AT THE OFFICES OF THE NEW RIVER COMPANY.

[*To face p.* 112.

rendered to the public by succeeding engineers, there has been no such cordial or graceful recognition of them by any succeeding monarch. The patent states that King James had made a baronet of Hugh Myddelton, of London, goldsmith, for the following reasons and considerations :—

"1. For bringing to the city of London, with excessive charge and greater difficulty, a new cutt or river of fresh water, to the great benefit and inestimable preservation thereof. 2. For gaining a very great and spacious quantity of land in Brading Haven, in the Isle of Wight, out of the bowells of the sea, and with bankes and pyles and most strange defensible and chargeable mountains, fortifying the same against the violence and fury of the waves. 3. For finding out, with a fortunate and prosperous skill, exceeding industry, and noe small charge, in the county of Cardigan, a royal and rych myne, from whence he hath extracted many silver plates which have been coyned in the Tower of London for current money of England." *

The King, however, did more than confer the title—he added to it a solid benefit in confirming the lease made to Sir Hugh by the Governor and Company of Mines Royal, "as a recompense for his industry in bringing a new river into London," waiving all claim to royalty upon the silver produced, although the Crown was entitled, according to the then interpretation of the law, to a payment on all gold and silver found in the lands of a subject; and it is certain that the lessee† who succeeded Sir Hugh did pay

* Harleian MS., No. 1507, Art. 40. (British Museum.)
† Subsequent to 1636, Thomas Bushell (who purchased the lease) paid 1000l. per annum to the King ; and some years after, in 1647, we find him agreeing to pay 2500l per annum to the Parliament. As a curious fact, we may here add that, under date Die

such royalty into the State Exchequer. It also appears from documents preserved amongst the State Papers, that large offers of royalty were actually made to the King at the very time that this handsome concession was granted to Sir Hugh.

The discovery of silver in the Welsh mountains doubtless caused much talk at the time, and, as in Australia and California now, there were many attempts made by lawless persons to encroach upon the diggings. On this, a royal proclamation was published, warning such persons against the consequences of their trespass, and orders were issued that summary proceedings should be taken against them. It appears that Sir Hugh and his partners continued to work the mines with profit for a period of about sixteen years, although it is stated that during most of that time, in consequence of the large quantity of water met with, little more than the upper surface could be got at. The water must, however, have been sufficiently kept under to enable so much ore eventually to be raised. Waller says an engine was employed at Cwmsymlog; and a tradition long existed among the neighbouring miners that there were two engines placed about the middle of the work. There were also several "levels" at Cwmsymlog, one of which is called to this day "Sir Hugh's Level."

The following rude cut, from Pettus' 'Fodinæ Regales,' may serve to give an idea of the manner

Sabbati, 14 August, 1641, Parliament granted an order or iicense to Thomas Bushell to dig turf on the King's wastes within the limits of Cardiganshire, for the purpose of smelting and refining the lead ores, &c., his predecessor (Myddelton) having used up almost all the wood growing in the neighbourhood of the mines.

in which the works of Cwmsymlog (facetiously styled by the author or his printer "Come-some-luck") were laid out:

Plan of Myddelton's Silver Mining Works at Cwmsymlog.

A. The old works of Myddelton and Bushell.
B. The round holes are the shafts of the mine.
C. Windlace to wind up ore from the shafts.
D. A new vein.
E. Sir H. Myddelton's adit.
F. A new adit.

G. Adits to drain works.
H. Myddelton's decayed chapel.
I. Old stamping-house.
K. The smelting mills, supposed six miles from the hill.
L. Unwrought ground.
M. The brook that divides the hill.
N. The stream which drives the mills.

From a statement made by Bushell to Parliament of the results of the working subsequent to 1636, it appears that the lead alone was worth above 5000*l*. a year, to which there was to be added the value of the silver—Bushell alleging, in his petition to Charles I., deposited in the State Paper-office,* that Sir Hugh had brought

* Dated 22nd October, 1636. The prayer of Bushell's petition to Charles I. is, that His Majesty will ratify his agreement with Lady Myddelton (by that time a widow) for the purchase of the residue of her lease.

"to the Minte theis 16 yeares of puer silver 100 poundes weekly." A ton of the lead ore is said to have yielded about a hundred ounces of silver, and the yield at one time was such that Myddelton's profits were alleged by Bushell to have amounted to at least two thousand pounds a month. There is no doubt, therefore, that Myddelton realised considerable profits by the working of his Welsh mines, and that towards the close of his useful life he was an eminently prosperous man.*

Successful as he had been in his enterprise, he was ready to acknowledge the Giver of all Good in the matter. He took an early opportunity of presenting a votive cup, manufactured by himself out of the Welsh silver, to the corporation of Denbigh, and another to the head of his family at Gwaenynog, in its immediate neighbourhood, both of which are still preserved. On the latter is inscribed "Mentem non munus— Omnia a Deo—Hugh Myddelton."

While conducting the mining operations, Sir Hugh resided at Lodge, now called Lodge Park, in the immediate neighbourhood of the mines. The house was the property of Sir John Pryse, of Gogerddan, whose son Richard, afterwards created a baronet, was married to Myddelton's daughter Hester. The house stood on the top of a beautifully wooded hill, overlooking the

---

* In these rampant 'days of joint-stock enterprise, we are not surprised to observe that a scheme has been set on foot to work the long-abandoned silver and lead mines of Sir Hugh Myddelton. With the greatly improved mining machinery of modern times, it is not unreasonable to anticipate a considerable measure of success from such an undertaking.

estuary of the Dovey and the great bog of Gors-
fochno, the view being bounded by picturesque
hills on the one hand and by the sea on the
other. Whilst residing here, on one of his visits
to the mines, a letter reached him from his cousin,
Sir John Wynn, of Gwydir, dated the 1st September, 1625, asking his assistance in an engineering project in which he was interested. This
was the reclamation of the large sandy marshes,
called Traeth-Mawr and Traeth-Bach, situated at
the junction of the counties of Caernarvon and
Merioneth, at the northern extremity of the bay
of Cardigan. Sir John, after hailing his good
cousin as "one of the great honours of the
nation," congratulated him on the great work
which he had performed in the Isle of Wight,
and added, "I may say to you what the Jews
said to Christ, We have heard of thy greate
workes done abroade, doe now somewhat in thine
own country." After describing the nature of the
land proposed to be reclaimed, Sir John declares
his willingness "to adventure a brace of hundred
pounds to joyne with Sir Hugh in the worke,"
and concludes by urging him to take a ride to
Traeth - Mawr, which was not above a day's
journey from where Sir Hugh was residing, and
afterwards to come on and see him at Gwydir
House, which was at most only another day's
journey or about twenty-five miles further to the
north-west of Traeth-Mawr. The following was
Sir Hugh's reply :—

"HONOURABLE SIR,

"I have received your kind letter. Few are the
things done by me ; for which I give God the glory. It may
please you to understand my first undertaking of public works

was amongst my owne kindred, within less than a myle of the place where I hadd my first being, 24 or 25 years since, in seekinge of coales for the town of Denbighe.

"Touching the drowned lands near your lyvinge, there are many things considerable therein. Iff to be gayned, which will hardlie be performed without great stones, which was plentiful at the Weight [Isle of Wight], as well as wood, and great sums of money to be spent, not hundreds, but thousands;* and first of all his Majesty's interest must be got. As for myself, I am grown into years, and full of business here at the mynes, the river at London, and other places, my weeklie charge being above 200*l.*; which maketh me verie unwillinge to undertake any other worke; and the least of theis, whether the drowned lands or mynes, requireth a whole man, with a large purse. Noble sir, my desire is great to see you, which should draw me a farr longer waie; yet such are my occasions at this tyme here, for the settlinge of this great worke, that I can hardlie be spared one howr in a daie. My wieff being also here, I cannot leave her in a strange place. Yet my love to publique works, and desire to see you (if God permit), maie another tyme drawe me into those parts. Soe with my heartie comendations I comit you and all your good desires to God.

"Your assured lovinge couzin to command,

"HUGH MYDDELTON.

"Lodge, Sept. 2nd, 1625."

---

* A long time passed before the attempt was made to reclaim the large tract of land at Traeth-Mawr; but after the lapse of two centuries it was undertaken by William Alexander Madocks, Esq., and accomplished in spite of many formidable difficulties. Two thousand acres of Penmorfa Marsh were first enclosed on the western side of the river, after which an embankment was constructed across the estuary, about a mile in length, by which 6000 additional acres were secured. The sums expended on the works are said to have exceeded 100,000*l.*; but the expenditure has proved productive, and the principal part of the reclaimed land is now under cultivation. Tremadoc, or Madock's Town, and Port Madoc, are two thriving towns, built by the proprietor on the estate thus won from the sea.

At the date of this letter Sir Hugh was an old
man of seventy, yet he still continued industriously
to apply himself to business affairs. Like most
men with whom work has become a habit, he
could not be idle, and active occupation seems to
have been necessary to his happiness. To the
close of his life we find him engaged in correspondence on various subjects—on mining, draining, and general affairs. When in London he
continued to occupy his house in Bassishaw-
street, where the goldsmith business was carried
on in his absence by his son William. He also
continued to maintain his pleasant country house
at Bush Hill, near Edmonton, which he occupied
when engaged on the engineering business of
the New River, near to which it was conveniently
situated.

At length all correspondence ceases, and the
busy hand and head of the old man find rest in
death. Sir Hugh died on the 10th of December,
1631, at the advanced age of seventy-six. In his
will, which he made on the 21st November, three
weeks before his death, when he was "sick in
bodie" but "strong in mind," for which he praised
God, he directed that he should be buried in the
church of St. Matthew, Friday-street, where he
had officiated as churchwarden, and where six of
his sons and five of his daughters had been
baptized. It had been his parish church, and was
hallowed in his memory by many associations of
family griefs as well as joys ; for there he had
buried several of his children in early life, amongst
others his two eldest-born sons. The church of
St. Matthew, however, has long since ceased to
exist, though its registers have been preserved :

it was destroyed in the great fire of 1666, and the monumental record of Sir Hugh's last resting-place perished in the common ruin.

The popular and oft-repeated story of Sir Hugh Myddelton having died in poverty and obscurity is only one of the numerous fables which have accumulated about his memory.* He left fair portions to all the children who survived him, and an ample provision to his widow.† His eldest son and heir, William, who succeeded to the baronetcy, inherited the estate at Ruthin, and afterwards married the daughter of Sir Thomas Harris, Baronet, of Shrewsbury. Elizabeth, the daughter

---

* The tradition still survives that Sir Hugh retired in his old age to the village of Kemberton, near Shiffnal, Salop, where he lived in great indigence under the assumed name of Raymond, and that he was there occasionally employed as a street paviour! The parish register is said to contain an entry of his burial on the 11th of March, 1702 ; by which date Hugh Myddelton, had he lived until then, would have been about 150 years old! The entry in the register was communicated by the rector of the parish in 1809 to the 'Gentleman's Magazine' (vol. lxxix., p. 795), but it is scarcely necessary to point out that it can have no reference whatever to the subject of this memoir.

† On the 24th June, 1632, Lady Myddelton memorialised the Common Council of London with reference to the loan of 3000l. advanced to Sir Hugh, which does not seem to have been repaid ; and more than two years later, on the 10th Oct., 1634, we find the Corporation allowed 1000l. of the amount, in consideration of the public benefit conferred on the city by Sir Hugh through the formation of the New River, and for the losses alleged to have been sustained by him through breaches in the water-pipes on the occasion of divers great fires, as well as for the "present comfort" of Lady Myddelton. It is to be inferred that the balance of the loan of 3000l. was then repaid. Lady Myddelton died at Bush Hill on the 19th July, 1643, aged sixty-three, and was interred in the chancel of Edmonton Church, Middlesex. On her monumental tablet it is stated that she was "the mother of fifteen children."

of Sir William, married John Grene, of Enfield,
clerk to the New River Company, and from her is
lineally descended the Rev. Henry Thomas Ella-
combe, M.A., rector of Clyst St. George, Devon,
who still holds two shares in the New River
Company, as trustee for the surviving descendants of Myddelton in his family. Sir Hugh left
to his two other sons, Henry and Simon,* besides
what he had already given them, one share each
in the New River Company (after the death of his
wife) and 400*l*. a-piece. His five daughters seem
to have been equally well provided for. Hester
was left 900*l*., the remainder of her portion of
1900*l*.; Jane having already had the same portion
on her marriage to Dr. Chamberlain, of London.
Elizabeth and Ann, like Henry and Simon, were
left a share each in the New River Company and
500*l*. a-piece. He bequeathed to his wife, Lady
Myddelton, the house at Bush Hill, Edmonton,
and the furniture in it, for use during her life,
with remainder to his youngest son Simon and
his heirs. He also left her all the "chains, rings,
jewels, pearls, bracelets, and gold buttons, which
she hath in her custody and useth to wear at
festivals, and the deep silver basin, spout pot,
maudlin cup, and small bowl;" as well as "the
keeping and wearing of the great jewel given to
him by the Lord Mayor and Aldermen of London,
and after her decease to such one of his sons as
she may think most worthy to wear and enjoy it."
By the same will Lady Myddelton was authorised
to dispose of her interest in the Cardiganshire

---

* Simon's son Hugh was created a Baronet, of Hackney,
Middlesex, in 1681. He married Dorothy, the daughter of Sir
William Oglander, of Nunwell, Baronet.

mines for her own benefit; and it afterwards appears, from documents in the State Paper Office,
that Thomas Bushell, "the great chymist," as he
was called, purchased it for 400*l.* cash down, and
400*l.* per annum during the continuance of her
grant, which had still twenty-five years to run
after her husband's death.

Besides these bequeathments, and the gifts of
land, money, and New River shares, which he had
made to his other children during his lifetime,
Sir Hugh left numerous other sums to relatives,
friends, and clerks; for instance, to Richard Newell
and Howell Jones, 30*l.* each, "to the end that the
former may continue his care in the works in the
Mines Royal, and the latter in the New River
water-works," where they were then respectively
employed. He also left an annuity of 20*l.* to
William Lewyn, who had been engaged in the
New River undertaking from its commencement.
Nor were his men and women servants neglected,
for he bequeathed to each of them a gift of money,
not forgetting "the boy in the kitchen," to whom
he left forty shillings. He remembered also the
poor of Henllan, near Denbigh, "the parish in
which he was born," leaving to them 20*l.*; a similar
sum to the poor of Denbigh, which he had represented in several successive Parliaments; and 5*l.*
to the parish of Amwell, in Hertfordshire. To the
Goldsmiths' Company, of which he had so long
been a member, he bequeathed a share in the
New River Company, for the benefit of the more
necessitous brethren of that guild, "especially to
such as shall be of his name, kindred, and county."*

* Several of the descendants of Sir Hugh Myddelton, when
reduced in circumstances, obtained assistance from this fund. It

Such was the life and such the end of Sir Hugh Myddelton, a man full of enterprise and resources, an energetic and untiring worker, a great conqueror of obstacles and difficulties, an honest and truly noble man, and one of the most distinguished benefactors the city of London has ever known.

has been stated, and often repeated, that Lady Myddelton, after her husband's death, became a pensioner of the Goldsmiths' Company, receiving from them 20*l.* a year. But this annuity was paid, not to the widow of the first Sir Hugh, but to the mother of the last Sir Hugh, more than a century later. The last who bore the title was an unworthy scion of this distinguished family. He could raise his mind no higher than the enjoyment of a rummer of ale ; and towards the end of his life existed upon a pension granted him by the New River Company. The statements so often published (and which, on more than one occasion, have brought poor persons up to town from Wales to make inquiries) as to an annuity of 100*l.* said to have been left by Sir Hugh and unclaimed for a century, and of an advertisement calling upon his descendants to apply for the sum of 10,000*l.*, alleged to be lying for them at the Bank of England, are altogether unfounded. No such annuity has been left, no such sum has accrued, and no such advertisement has appeared.

CHAPTER V.

CAPTAIN PERRY—STOPPAGE OF DAGENHAM BREACH.

ALTHOUGH the cutting of the New River involved
a great deal of labour, and was attended with
considerable cost, it was not a work that would
now be regarded as of any importance in an
engineering point of view. It was, nevertheless,
one of the greatest undertakings of the kind that
had at that time been attempted in England; and
it is most probable that, but for the persevering
energy of Myddelton and the powerful support of
the King, the New River enterprise would have
failed. As it was, a hundred years passed before
another engineering work of equal importance
was attempted, and then it was necessity, and not
enterprise, that occasioned it.

We have, in a previous chapter, referred to the
artificial embankment of the Thames, almost from
Richmond to the sea, by which a large extent of
fertile land is protected from inundation along
both banks of the river. The banks first raised
seemed to have been in many places of insufficient
strength; and when a strong north-easterly wind
blew down the North Sea, and the waters became
pent up in that narrow part of it lying between
the Belgian and the English coasts,—and especially
when this occurred at a time of the highest spring

tides,—the strength of the river embankments became severely tested throughout their entire length, and breaches often took place, occasioning destructive inundations.

Down to the end of the seventeenth century scarcely a season passed without some such accident occurring. There were frequent burstings of the banks on the south side between London Bridge and Greenwich, the district of Bermondsey, then green fields, being especially liable to be submerged. Commissions were appointed on such occasions with full powers to distrain for rates, and to impress labourers in order that the requisite repairs might at once be carried out. In some cases the waters for a long time held their ground, and refused to be driven back. Thus, in the reign of Henry VIII., the marshes of Plumstead and Lesnes, now used as a practising ground by the Woolwich garrison, were completely drowned by the waters which had burst through Erith Breach, and for a long time all measures taken to reclaim them proved ineffectual. There were also frequent inundations of the Combe Marshes, lying on the east of the royal palace at Greenwich.

But the most destructive inundations occurred on the north bank of the Thames. Thus, in the year 1676, a serious breach took place at Lime-house, when many houses were swept away, and it was with the greatest difficulty that the waters could be banked out again. The wonder is, that sweeping, as the new current did, over the Isle of Dogs, in the direction of Wapping, and in the line of the present West India Docks, the channel of the river was not then permanently altered. But Deptford was already established as a royal dockyard, and probably the diversion of the river would have inflicted as much local injury, judging by comparison, as it unquestionably would do at the present day. The breach was accordingly stemmed, and the course of the river held in its ancient channel by Deptford and Greenwich. Another destructive inundation shortly after occurred through a breach made in the embankment of the West Thurrock Marshes, in what is called the Long Reach, nearly opposite Greenhithe, where the lands remained under water for seven years, and it was with much difficulty that the breach could be closed.

But the most destructive and obstinate of all the breaches was that made in the north bank a little to the south of the village of Dagenham, in Essex, by which the whole of the Dagenham and Havering Levels lay drowned at every tide. A similar breach had occurred in 1621, which Vermuyden, the Dutch engineer, succeeded in stopping; and at the same time he embanked or "inned" the whole of Dagenham Creek, through which the little rivulet flowing past the village of that name found its way to the Thames. Across the mouth of this rivulet Vermuyden had erected a sluice, of the nature of a "clow," being a strong gate suspended by hinges, which opened to admit of the egress of the inland waters at low tide, and closed against the entrance of the Thames when the tide rose. It happened, however, that a heavy inland flood, and an unusually high spring tide, occurred simultaneously during the prevalence of a strong north-easterly wind, in the year 1707; when the united force of the waters meeting from both directions blew up the sluice, the repairs of which had been neglected, and in a very short time nearly the whole area of the above Levels was covered by the waters of the Thames.

At first the gap was so slight as to have been easily closed, being only from 14 to 16 feet wide. But no measures having been taken to stop it, the tide ran in and out for several years, every tide wearing the channel deeper, and rendering the stoppage of the breach more difficult. At length the channel was found upwards of 30 feet deep at low water, and about 100 feet wide, a lake more than a mile and a half in extent having by this time been formed inside the line of the river embankment. Above a thousand acres of rich lands were spoiled for all useful purposes, and by the scouring of the waters out and in at every tide, the soil of about a hundred and twenty acres was completely washed away. It was carried into the channel of the Thames, and formed a bank of about a mile in length, reaching half-way across the river. This state of things could not be allowed to continue, for the navigation of the stream was seriously interrupted by the obstruction, and there was no knowing where the mischief would stop.

Various futile attempts were made by the adjoining landowners to stem the breach. They filled old ships with chalk and stones, and had them scuttled and sunk in the deepest places, throwing in baskets of chalk and earth outside them, together with bundles of straw and hay to stop up the interstices; but when the full tide rose, it washed them away like so many chips, and the opening was again driven clean through. Then the expedient was tried of sinking into the hole gigantic boxes made expressly for the purpose, fitted tightly together, and filled with chalk. Power was obtained to lay an embargo on the cargoes of chalk and ballast contained in passing ships, for the purpose of filling these boxes, as well as damming up the gap; and as many as from ten to fifteen freights of chalk a day were thrown in, but still without effect.

One day when the tide was on the turn, the force of the water lifted one of the monster trunks sheer up from the bottom, when it toppled round, the lid opened, out fell the chalk, and, righting again, the immense box floated out into the stream and down the river. One of the landowners interested in the stoppage ran along the bank, and shouted out at the top of his voice, "Stop her! stop her!" But the unwieldy object being under no guidance was carried down stream towards the shipping lying at Gravesend, where its unusual appearance, standing so high out of the water, excited great alarm amongst the sailors. The empty trunk, however, floated safely past, down the river, until it reached the Nore, where it stranded upon a sandbank.

The Government next lent the undertakers an old royal ship called the *Lion*, for the purpose of being sunk in the breach, which was done, with two other ships; but the *Lion* was broken in pieces by a single tide, and at the very next ebb not a vestige of her was to be seen. No matter what was sunk, the force of the water at high tide bored through underneath the obstacle, and only served to deepen the breach. After the destruction of the *Lion*, the channel was found deepened to 50 feet at low water, at the very place where she had been sunk.

All this had been but tinkering at the breach, and every measure that had been adopted merely proved the incompetency of the undertakers. The obstruction to the navigation through the deposit of earth and sand in the river being still on the increase, an Act was passed in 1714, after the bank had been open for a period of seven years, giving powers for its repair at the public expense. But it is an indication of the very low state of engineering ability in the kingdom at the time, that several more years passed before the measures taken with this object were crowned with success, and the opening was only closed after a fresh succession of failures.

The works were first let to one Boswell, a contractor. He proceeded very much after the method which had already failed, sinking two rows of caissons or chests across the breach, but provided with sluices for the purpose of shutting off the inroads of the tide. All his contrivances, however, failed to make the opening watertight; and his chests were blown up again and again. Then he tried pontoons of ships, which he loaded and sunk in the opening; but the force of the tide, as before, rushed under and around them, and broke them all to pieces, the only result being to make the gap in the bank considerably wider and deeper than he found it. Boswell at length abandoned all further attempts to close it, after suffering a heavy loss; and the engineering skill of England seemed likely to be completely baffled by this hole in a river's bank.

The competent man was, however, at length found in Captain Perry, who had just returned from Russia, where, having been able to find no

I.   K

suitable employment for his abilities in his own
country, he had for some time been employed by
the Czar Peter in carrying on extensive engineering
works.

John Perry was born at Rodborough, in Gloucestershire, in 1669, and spent the early part of his life
at sea. In 1693 we find him a lieutenant on board
the royal ship the *Montague*. The vessel having put
into harbour at Portsmouth to be refitted, Perry
is said to have displayed considerable mechanical
skill in contriving an engine for throwing out a
large quantity of water from deep sluices (probably for purposes of dry docking) in a very short
space of time. The *Montague* having been repaired, went to sea, and was shortly after lost.
As the English navy had suffered greatly during
the same year, partly by mismanagement, and
partly by treachery, the Government was in a
very bad temper, and Perry was tried for alleged
misconduct. The result was, that he was sen-
tenced to pay a fine of 1000*l*., and to undergo ten
years' imprisonment in the Marshalsea.

This sentence must, however, have been subsequently mitigated, for we find him in 1695 pub-
lishing a "Regulation for Seamen," with a view
to the more effectual manning of the English navy ;
and in 1698 the Marquis of Caermarthen and others
recommended him to the notice of the Czar Peter,
then resident in England, by whom he was invited
to go out to Russia, to superintend the establishment of a royal fleet, and the execution of several
gigantic works then contemplated for the purpose
of opening up the resources of that empire. Perry
was engaged by the Czar at a salary of 300*l*. a
year, and shortly after accompanied him to Holland,

thence proceeding to Moscow, to enter upon the business of his office.

One of the Czar's grand designs was to open up a system of inland navigation to connect his new city of St. Petersburg with the Caspian Sea, and also to place Moscow upon another line, by forming a canal between the Don and the Volga. In 1698 the works had been begun by one Colonel Breckell, a German officer in the Czar's service. But though a good military engineer, it turned out that he knew nothing of canal making; for the first sluice which he constructed was immediately blown up. The water, when let in, forced itself under the foundations of the work, and the six months' labour of several thousand workmen was destroyed in a night. The Colonel, having a due regard for his personal safety, at once fled the country in the disguise of a servant, and was never after heard of. Captain Perry entered upon this luckless gentleman's office, and forthwith proceeded to survey the work he had begun, some seventy-five miles beyond Moscow. Perry had a vast number of labourers at his disposal, but they were altogether unskilled, and therefore comparatively useless. His orders were to have no fewer than 30,000 men at work, though he seldom had more than from 10,000 to 15,000; but one-twentieth the number of skilled labourers would have better served his purpose. He had many difficulties to contend with. The local nobility or boyars were strongly opposed to the undertaking, declaring it to be impossible; and their observation was, that God had made the rivers to flow one way, and it was presumption in man to think of attempting to turn them in another.

K 2

Shortly after the Czar had returned to his dominions, he got involved in war with Sweden, and was defeated by Charles XII. at the battle of Narva, in 1701. Although the Don and Volga Canal was by this time half-dug, and many of the requisite sluices were finished, the Czar sent orders to Perry to let the works stand, and attend upon him immediately at St. Petersburg. Leaving one of his assistants to take charge of the work in hand, Perry waited upon his royal employer, who had a great new design on foot of an altogether different character. This was the formation of a royal dockyard on one of the southern rivers of Russia, where Peter contemplated building a fleet of war-ships, wherewith to act against the Turks in the Black Sea. Perry immediately entered upon the office to which he was appointed, of Comptroller of Russian Maritime Works, and proceeded to carry out the new project. The site of the Royal Dockyard was fixed at Veronize on the Don, where he was occupied for several years, with a vast number of workmen under him, in building a dockyard, with storehouses, ship-sheds, and workshops. He also laid down and superintended the construction of numerous vessels, one of them of eighty guns : the slips on which he built them are said to have been very ingeniously contrived.

The creation of this dockyard was far advanced when he received a fresh command to undertake the survey of a canal to connect St. Petersburg with the Volga, to enable provisions, timber, and building materials to flow freely to the capital from the interior of the empire. Perry surveyed three several routes, recommending the adoption of that through Lakes Ladoga and Onega ; and the works were forthwith begun under his direction. Before they were completed, however, he had left Russia, never to return. During the whole of his stay in the kingdom he had been unable to get paid for his work. His applications for his stipulated salary were put off with excuses from year to year. Proceedings in the courts of law were out of the question in such a country; he could only dun the Czar and his ministers; and at length his arrears had become so great, and his necessities so urgent, that he could no longer endure his position, and threatened to quit the Czar's service. It came to his ears that the Czar had threatened on his part, that if he did, he would have Perry's head; and the engineer immediately took refuge at the house of the British minister, who shortly after contrived to get him conveyed safely out of the country, but without being paid. He returned to England in 1712, as poor as he had left it, though he had so largely contributed to create the navy of Russia, and to lay the foundations of its afterwards splendid system of inland navigation.

It will be remembered that all attempts made to stop the breach at Dagenham had thus far proved ineffectual; and it threatened to bid defiance to the engineering talent of England. Perry seemed to be one of those men who delight in difficult undertakings, and he no sooner heard of the work than he displayed an eager desire to enter upon it. He went to look at the breach shortly after his return, and gave in a tender with a plan for its repair; but on Boswell's being accepted, which was the lowest, he held back until that contractor had tried his best, and failed. The way was now clear for Perry, and again he offered

to stop the breach and execute the necessary works
for the sum of 25,000*l.** His offer was this time
accepted, and operations were begun early in 1715.
The opening was now of great width and depth,
and a lake had been formed on the land from 400
to 500 feet broad in some places, and extending
nearly 2 miles in length. Perry's plan of operations may be briefly explained with the aid of his
own map.

In the first place he sought to relieve the tremendous pressure of the waters against the breach
at high tide, by making other openings in the bank
through which they might more easily flow into
and out of the inland lake, without having exclusively to pass through the gap which it was his
object to stop. He accordingly had two openings,
protected by strong sluices, made in the bank a
little below the breach, and when these had been
opened and were in action he proceeded to stop
the breach itself. He began by driving in a row of
strong timber piles across the channel; and they
were dovetailed one into the other so as to render
them almost impervious to water. The heads
of the piles were not more than from eighteen
inches to two feet above low water mark, so that
in driving them little or no difficulty would be
experienced from the current of ebb or flood.
" Forty feet from this central row of sheeting piles,
was constructed on each side, a sort of low coffer-
dam-like structure, variously stated as 18 or 20
feet broad, formed of vertical piles and horizontal
boarding, and filled with chalk, to prevent the toe
of the future embankment from spreading. On

* Boswell's price had been 16,300*l.*, and he undertook to do the
work in fifteen months.

the outside of these foot-wharfs, as Perry calls them, a wall of chalk rubble was made, as a further security. The dam itself was composed entirely

Perry's Plan of Dagenham Breach.

A.—The Dam whereby the Breach was stopped.
B —The site of Boswell's works.
C.—The site of the Landowners' works.
D.—The site of Perry's Sluices.
E.—The site of Boswell's Sluices.
F.—A Dam and Sluice made for recovery of the Meadows shortly after the Breach had occurred.

G.—Small Sluice for drainage of the land waters.
HH.—The dotted line represents the extent of the inundation caused by the Breach.
I.—Places where stags' horns were dug up.
K.—Parallel lines, showing the depth at low water at every 60 yards distance from the shore.

of clayey earth, in layers about 3 feet in height, and scarcements or steps of about 7 feet; and in the course of its erection, care was taken always to shut the sluices already mentioned when, at each successive ebb-tide, the level of the backwater fell to the level of the top of the work in progress. In this way there was at no time a higher face for the water of the rising tide to flow over. In fact the unfinished embankment held in the water, over the land it was intended to lay dry, at a depth corresponding to its gradual progress, until finally, when the bank was above highwater line, it was discharged by the sluices, and never re-admitted." *

Scarcely had Perry begun the work, and proceeded so far as to exhibit his general design, than Boswell, the former contractor, presented a petition to Parliament against the engineer being allowed to go on, alleging that his scheme was utterly impracticable. The work being of great importance, and executed at the public expense, a Parliamentary Committee was appointed, when Perry was called before them and examined fully as to the details. His answers were so explicit, and, on the whole, so satisfactory, that at the close of the examination one of the members thus spoke the sense of the Committee :—" You have answered us like an artist, and like a workman ; and it is not only the scheme, but the man, that we recommend."

Perry was then allowed to proceed, and the work went steadily forward. About three hundred men were employed in stopping the breach, and it occupied them about five years to accomplish it. "Perry was proceeding steadily with the dam, which was constructed by successive scarcements

* Description of Perry's work by S. Downing, C.E., in *Practical Mechanics' Journal*, May, 1864.

about 7 feet broad and 3 feet high; these being
supported by piles and planking on the side, and
protected by layers of reeds on the top, had been
able to resist the action of the tide when it came
on. In this manner he was advancing to completion, when one of his assistants proposed to
the parties who had advanced Captain Perry the
necessary capital, to set all hands to work at
neap tides, and form a narrow wall of earth, un-

Dagenham Lake.

protected by reeds or planking, and build it so
rapidly as to get it above the level of the
springs before they should come on, and thus at
once exclude the tides from the level. Unfortunately, the next spring-tide rose to an unexpected
height under the influence of a storm from the

north-west, and over-topped this narrow dam by about six inches, although Perry used the greatest energy, and heightened the wall of earth by piles and boarding set on edge on the top ; but all in vain : the water poured over it, and in the course of *two hours* the whole dam was swept away, and the dovetailed piles laid bare. This accident was repaired in the winter months, and in June, 1718, the tide was again turned out of the levels ; but in September of the same year the dam gave way again, and this time with far greater injury to the work, as upwards of 100 feet of the dovetailed piles were torn up and carried away. In one place there was about 20 feet greater depth than before the work was begun. The third dam was completed on the 18th June, 1719, about fourteen years after the accident first occurred." Thus the opening was at length effectually stopped, and the water drained away by the sluices, leaving the extensive inland lake, which is to this day used by the Londoners as a place for fishing and aquatic recreation.*

A good idea of the formidable character of the embankments extending along the Thames may be obtained by a visit to this place. Standing on the top of the bank, which is from 40 to 50 feet above the river level at low water,† we see on the one side the Thames, with its shipping passing and

---

* It has recently been proposed to convert the lake into extensive docks, connected with London by the Tilbury Railway ; and a Company has been formed for the purpose of carrying out the enterprise.

† The banks themselves are from 17 to 25 feet high in the neighbourhood of Dagenham, and from 25 to 30 feet wide at the base. The marks of the old breach are still easily traceable.

repassing, high above the inland level when the tide is up, with the still lake of Dagenham and the far extending flats on the other. Looking from the lower level on these strong banks extending along the stream as far as the eye can reach, we can only see the masts of sailing ships and the funnels of large steamers leaving behind them long trails of murky smoke,—at once giving an idea of the gigantic traffic that flows along this great water

The Thames from Dagenham Bank, looking up the River.

highway, and the enormous labour which it has cost to bank up the lands and confine the river within its present artificial limits. We do not exaggerate when we state that these formidable embankments, winding along the river side, up creeks and tributary streams, round islands and about marshes, from London to the mouth of the Thames, are not less than 300 miles in extent.

It is to be regretted that Perry gained nothing but fame by his great work. The expense of stopping the breach far exceeded his original estimate; he required more materials than he had calculated upon; and frequent strikes amongst his workmen for advances of wages greatly increased the total cost. These circumstances seem to have been taken into account by the Government in settling with the engineer, and a grant of 15,000*l.* was voted to him in consideration of his extra outlay. The landowners interested also made him a present of a sum of 1000*l.* But even then he was left a loser; and although the public were so largely benefited by the success of the work, which restored the navigation of the river, and enabled the adjoining proprietors again to reclaim for purposes of agriculture the drowned lands within the embankment, the engineer did not really receive a farthing's remuneration for his five years' anxiety and labour.

After this period Perry seems to have been employed on harbour works,—more particularly at Rye and Dover; but none of these were of great importance, the enterprise of the country being as yet dormant, and its available capital for public undertakings comparatively limited. It appears from the Corporation Records of Rye, that in 1724 he was appointed engineer to the proposed new harbour-works there. The port had become very much silted up, and for the purpose of restoring the navigation it was designed to cut a new channel, with two pier-heads, to form an entrance to the harbour. The plan further included a large stone sluice and draw-bridge, with gates, across the new channel, about a quarter of a mile within the pier-heads; a wharf constructed of timber along the two sides of the channel, up to the sluice; together with other well-designed improvements. But the works had scarcely been begun before the Commissioners displayed a strong disposition to job, one of them withdrawing for the purpose of supplying the stone and timber required for the new works at excessive prices, and others forming what was called "the family compact," or a secret arrangement for dividing the spoil amongst them. The plan of Perry was not fully carried out; and though the pier-heads and stone sluice were built, the most important part of the work, the cutting of the new channel, was only partly executed, when the undertaking was suspended for want of funds.

From that time forward, Perry's engineering ability was very much confined to making reports as to what things should be done, rather than in being employed to do them. In 1727 he published his 'Proposals for Draining the Fens in Lincolnshire;' and he seems to have been employed there as well as in Hatfield Level, where "Perry's Drain" still marks one of his works. He was acting as engineer for the adventurers who undertook the drainage of Deeping Fen, in 1732, when he was taken ill and died at Spalding, in the sixty-third year of his age. He lies buried in the churchyard of that town; and the tombstone placed over his grave bears the following inscription :—

To the Memory of

IOHN PERRY Esqr; in 1693

Commander of His Maiesty King Will^{m,s}
Ship the Cignet ; second Son of Saml Perry
of Rodborough in Gloucestershire Gent & of
Sarah his Wife ; Daughter of Sir Thos Nott; K^r
He was several Years Comptroller of the
Maritime works to Czar Peter in Russia &
on his Return home was Employed by y^e
Parliament to stop Dagenham Breach which
he Effected and thereby Preserved the
Navigation of the River of Thames and
Rescued many Private Familys from Ruin
he after departed this Life in this Town &
was here Interred February 13 ; 1732 Aged
63 Years
This stone was placed over him by the
Order of William Perry of Penthurst in
Kent Esqr his Kindsman and Heir Male

CHAPTER VI.

IN the preceding memoirs of Vermuyden and Perry, we have found a vigorous contest carried on against the powers of water, the chief object of the engineers being to dam it back by embankments, or to drain it off by cuts and sluices; whilst in the case of Myddelton, on the other hand, we find his chief concern to have been to collect all the water within his reach, and lead it by conduit and aqueduct for the supply of the thirsting metropolis. The engineer whose history we are now about to relate dealt with water in like manner to Myddelton, but on a much larger scale; directing it into extensive artificial canals, for use as the means of communication between various towns and districts.

Down to the middle of last century, the trade and commerce of England were comparatively insignificant. This is sufficiently clear from the wretched state of our road and river communication about that time; for it is well understood that without the ready means of transporting commodities from place to place, either by land or water, commerce is impossible. But the roads of England were then about the worst in Europe, and usually impassable for vehicles during the greater part of

the year.*   Corn, wool, and such like articles, were
sent to market on horses or bullocks' backs; and
manure was carried to the field, and fuel conveyed
from the forest or the bog, in the same way.   The
only coal used in the inland southern counties was
carried on horseback in sacks for the supply of the
blacksmiths' forges.   The food of London was principally brought from the surrounding country in
panniers.   The little merchandise transported from
place to place was mostly of a light description,—
the cloths of the West of England, the buttons
of Birmingham and Macclesfield, the baizes of
Norwich, the cutlery of Sheffield, and the tapes,
coatings, and fustians of Manchester.

Articles imported from abroad were in like
manner conveyed inland by pack-horse or waggon;
and it was then cheaper to bring most kinds of
foreign wares from remote parts to London by
sea than to convey them from the inland parts of
England to London by road.   Thus, two centuries
since, the freight of merchandise from Lisbon to
London was no greater than the land carriage of
the same articles from Norwich to London; and
from Amsterdam or Rotterdam the expense of
conveyance was very much less.   It cost from
7l. to 9l. to convey a ton of goods from Birmingham to London, and 13l. from Leeds to London.
It will readily be understood that rates such as
these were altogether prohibitory as regarded
many of the articles now entering largely into
the consumption of the great body of the people.
Things now considered necessaries of life, in

* For a full account of Old Roads and Travelling in England,
we must refer the reader to 'Lives of the Engineers,' vol. iii.
pp. 1–89.

daily common use, were then regarded as luxuries, obtainable only by the rich. The manufacture of pottery was as yet of the rudest kind. Vessels of wood, of pewter, and even of leather, formed the principal part of the household and table utensils of genteel and opulent families ; and we long continued to import our cloths, our linen, our glass, our "Delph" ware, our cutlery, our paper, and even our hats, from France, Spain, Germany, Flanders, and Holland. Indeed, so long as corn, fuel, wool, iron, and manufactured articles had to be transported on horseback, or in rude waggons dragged over still ruder roads by horses or oxen, it is clear that trade and commerce could make but little progress. The cost of transport of the raw materials required for food, manufactures, and domestic consumption, must necessarily have formed so large an item as to have in a great measure precluded their use ; and before they could be made to enter largely into the general consumption, it was absolutely necessary that greater facilities should be provided for their transport.

England was not, however, like many other countries less favourably circumstanced, necessarily dependent solely upon roads for the means of transport, but possessed natural water communications, and the means of improving and extending them to an almost indefinite extent. She was provided with convenient natural havens situated on the margin of the world's great highway, the ocean, and had the advantage of fine tidal rivers, up which fleets of ships might be lifted at every tide into almost the heart of the land. Very little had as yet been done to take advantage of this great natural water power, and to extend navigation inland either by improving the rivers which might be made navigable, or by means of artificial canals, as had been done in Holland, France, and even Russia, by which those countries had in some parts been rendered in a great measure independent of roads.

It is true, public attention had from time to time been directed to the improvement of rivers and the cutting of canals, but excepting a few isolated attempts, little had been done towards carrying the numerous suggested plans in different parts of the country into effect. If we except some of the wider drains in the Fens, which were in certain cases made available for purposes of navigation, though to a very limited extent, the first canal was that constructed by John Trew, at Exeter, in 1566. In early times the tide carried vessels up to that city, but the Countess of Devon took the opportunity of revenging herself upon the citizens for some affront they had offered to her, by erecting a weir across the Exe at Topsham in 1284, which had the effect of closing the river to sea-going vessels. This continued until the reign of Henry VIII., when authority was granted by Parliament to cut a canal about three miles in length along the west side of the river, from Exeter to Topsham. The work was executed by Trew, and it is a curious circumstance that it contained the first lock constructed in England, —though locks are said to have been used in the Brenta in 1488, and were shortly after adopted in the Milan canals. John Trew was a native of Glamorganshire; and though he must have been a man of skill and enterprise, like many other projectors of improvements and benefactors of mankind, he seems to have realised only loss and mortification by his work. In consequence of an alleged failure on his part in carrying out the agreement for executing the canal, the Mayor and Chamber of the city disputed his claims, and he became involved in ruinous litigation. In a letter written by him to Lord Burleigh, in which he relates his suit against the Chamber of Exeter, Trew draws a sad picture of the state to which he was reduced. "The varyablenes of men," says he, "and the great injury done unto me, brought me in such case that I wyshed my credetours sattisfyd and I away from earth: what becom may of my poor wyf and children, who lye in great mysery, for that I have spent all." * He then proceeded to recount "the things whearin God hath given (him) exsperyance;" relating chiefly to mining operations, and various branches of civil and even military engineering. It is satisfactory to add that in 1573 the harassing suit was brought to a conclusion, and Trew granted the Corporation a release on their agreeing to pay him a sum of 224l., and thirty pounds a year for life.†

In the reign of James I. several Acts of Parliament were passed, giving powers to improve rivers, so as to facilitate the passage of boats and barges

* Lansdowne MS. cvii. art. 73.

† The Lansdowne MS. xxxi., art. 74, sets forth certain " Reasons against the proceedings of John Trew in the works of Dover Haven, 1581." It appears from Lysons's History of Dover, that Trew was engaged as harbour-engineer at ten shillings a-day wages ; but the corporation, thinking that he was inclined to protract the work on which he was engaged, summarily dismissed him. This is the last that we hear of John Trew.

carrying merchandise. Thus, in 1623, Sir Hugh
Myddelton was engaged upon a Committee on a
bill then under consideration "for the making of
the river of Thames navigable to Oxford." In the
same year Taylor, the water poet, pointed out to
the inhabitants of Salisbury that their city might
be effectually relieved of its poor by having their
river made navigable from thence to Christchurch.
The progress of improvement, however, must have
been slow; as urgent appeals, on the same subject,
continued to be addressed to Parliament and the
public for a century later.

In 1656 we find one Francis Mathew addressing
Cromwell and his Parliament on the immense
advantage of opening up a water-communication
between London and Bristol. But he only proposed to make the rivers Isis and Avon navigable
to their sources, and then either to connect their
heads by means of a short sasse or canal of about
three miles across the intervening ridge of country,
or to form a fair stone causeway between the heads
of the two rivers, across which horses or carts
might carry produce between the one and the
other. His object, it will be observed, was mainly
the opening up of the existing rivers; "and not,"
he says, "to have the old channel of any river to
be forsaken for a shorter passage." Mathew fully
recognised the formidable character of his project,
and considered it quite beyond the range of private
enterprise, whether of individuals or of any corporation, to undertake it; but he ventured to think
that it might not be too much for the power of
the State to construct the three miles of canal and
carry out the other improvements suggested by
him, with a reasonable prospect of success. The

scheme was, however, too bold for Mathew's time, and a century elapsed before another canal was made in England. A few years later, in 1677, a curious work was published by Andrew Yarranton,* in which he pointed out what the Dutch had accomplished by means of inland navigation, and what England ought to do as the best means of excelling the Dutch without fighting them. The main feature of his scheme was the improvement of our rivers so as to render them navigable and the inland country thus more readily accessible to commerce. For, in England, said he, there are large rivers well situated for trade, great woods, good wool and large beasts, with plenty of iron stone, and pit coals, with lands fit to bear flax, and with mines of tin and lead; and besides all these things in it, England has a good air. But to make these advantages available, the country, he held, must be opened up by navigation. First of all, he proposed that the Thames should be improved to Oxford, and connected with the Severn by the Avon to Bristol—these two rivers, he insisted, being the master rivers of England. When this has been done, says Mr. Yarranton, all the great and heavy carriage from Cheshire, all Wales, Shropshire, Staffordshire, and Bristol, will be carried to London and recarried back to the great towns, especially in the winter time, at half the rates they now pay, which will much promote and advance manufactures in the counties and places above named. "If I were a doctor," he says, "and could read a Lecture of the Circulation of the Blood, I should by that awaken all the

* 'England's Improvement by Sea and Land.' London, 1677.

City: For London is as the Heart is in the Body,
and the great Rivers are as its Veins; let them
be stopt, there will then be great danger either
of death, or else such Veins will apply themselves
to feed some other part of the Body, which it was
not properly intended for: For I tell you, Trade
will creep and steal away from any place, provided she may be better treated elsewhere." But
he goes on—"I hear some say, You projected the
making Navigable the River Stoure in Worcestershire: what is the reason it was not finished? I
say it was my projection, and I will tell you the
reason it was not finished. The River Stoure and
some other Rivers were granted by an Act of
Parliament to certain Persons of Honour, and
some progress was made in the work; but within
a small while after the Act passed it was let fall
again. But it being a brat of my own, I was not
willing it should be Abortive; therefore I made
offers to perfect it, leaving a third part of the
Inheritance to me and my heirs for ever, and we
came to an agreement. Upon which I fell on,
and made it compleatly Navigable from Sturbridge
to Kederminster; and carried down many hundred
Tuns of Coales, and laid out near one thousand
pounds, and then it was obstructed for Want of
Money, which by Contract was to be paid."

There is no question that this "want of money"
was the secret of the little progress made in the
improvement of the internal communications of
the country, as well as the cause of the backward
state of industry generally. England was then
possessed of little capital and less spirit, and hence
the miserable poverty, starvation, and beggary
which prevailed to a great extent amongst the

lower classes of society at the time when Mr. Yarranton wrote, and which he so often refers to in the course of his book. For the same reason most of the early Acts of Parliament for the improvement of navigable rivers remained a dead letter: there was not money enough to carry them out, modest though the projects usually were. Among the few schemes which were actually carried out about the beginning of the eighteenth century, was the opening up of the navigation of the rivers Aire and Calder, in Yorkshire. Though a work of no great difficulty, Thoresby speaks of it in his diary as one of vast magnitude. It was, however, of much utility, and gave no little impetus to the trade of that important district.

It was, indeed, natural that the demand for improvements in inland navigation should arise in those quarters where the communications were the most imperfect and where good communications were most needed, namely, in the manufacturing districts of the north of England. On the western side of the island Liverpool was then rising in importance, and the necessity became urgent for opening up its water communications with the interior. By the assistance of the tide, vessels were enabled to reach as high up the Mersey as Warrington; but there they were stopped by the shallows, which it was necessary to remove to enable them to reach Manchester and the adjacent districts. Accordingly, in 1720, an Act was obtained empowering certain persons to take steps to make navigable the rivers Mersey and Irwell from Liverpool to Manchester. This was effected by the usual contrivance of wears, locks, and flushes, and a considerable improvement in the navigation was thereby effected. Acts were also passed for the improvement of the Weaver navigation, the Douglas navigation, and the Sankey navigation, all in the same neighbourhood; and the works carried out proved of much service to the district.

But these improvements, it will be observed, were principally confined to clearing out the channels of existing rivers, and did not contemplate the making of new and direct navigable cuts between important towns or districts. It was not until about the middle of last century that English enterprise was fairly awakened to the necessity of carrying out a system of artificial canals throughout the kingdom; and from the time when canals began to be made, it will be found that the industry of the nation made a sudden start forward. Abroad, monarchs had stimulated like undertakings, and drawn largely on the public resources for the purpose of carrying them into effect; but in England such projects are usually left to private enterprise, which follows rather than anticipates the public wants. In the upshot, however, the English system, as it may be termed—which is the outgrowth in a great measure of individual energy—does not prove the least efficient; for we shall find that the English canals, like the English railways, were eventually executed with a skill, despatch, and completeness, which imperial enterprise, backed by the resources of great states, was unable to surpass or even to equal. How the first English canals were made, how they prospered, and how the system extended, will appear from the following biography of James Brindley, the father of canal engineering in England.

JAMES BRINDLEY.

[*To face p.* 153.

LIFE OF JAMES BRINDLEY.

BRINDLEY'S NATIVE DISTRICT.

LIFE OF JAMES BRINDLEY.

CHAPTER I.

In the third year of the reign of George I., whilst the British Government were occupied in extinguishing the embers of the Jacobite rebellion which had occurred in the preceding year, the first English canal engineer was born in a remote hamlet in the High Peak of Derby.

The nearest town of any importance was Macclesfield, where a considerable number of persons were employed about the middle of last century in making wrought buttons in silk, mohair, and twist. The articles were sold throughout the country by pedestrian hawkers, most of whom had squatted on the waste lands called "The Flash," from a hamlet of that name situated between Buxton, Leek, and Macclesfield. They were notorious for their half-barbarous manners, and brutal pastimes. Such was the district, and such the population, in the neighbourhood of which our hero was born.

James Brindley first saw the light in a humble cottage standing about midway between the hamlet of Great Rocks, and that of Tunstead, in the liberty of Thornsett, some three miles to the north-east of Buxton. The house in which he was born, in

the year 1716, has long since fallen to ruins—the
Brindley family having been its last occupants.
The walls stood for some time after the roof had
fallen in, and at length the materials were removed
to build cowhouses ; but in the middle of the ruin
there grew up a young ash tree, forcing up one of
the flags of the cottage floor.   It looked so healthy
and thriving a plant, that the labourer employed to

Brindley's Croft.*

remove the stones for the purpose of forming the
pathway to the neighbouring farm-house, spared
the seedling, and it grew up into the large and

* The site of the Croft is very elevated, and commands an
extensive view as far as Topley Pike, between Bakewell and
Buxton, at the top of what is called the Long Hill.   Topley Pike
is behind the spectator in looking at the Croft in the above aspect.
The rising ground behind the ash tree is called Wormhill Common,
though now enclosed.   The old road from Buxton to Tideswell
skirts the front of the rising ground.

flourishing tree, six feet nine inches in girth, standing in the middle of the Croft, and now known as "Brindley's Tree." This ash tree is Nature's own memorial of the birth-place of the engineer, and it is the only one as yet erected in commemoration of his genius.

Although the enclosure is called Brindley's Croft, this name was only given to it of late years by its tenant, in memory of the engineer who was born there. The statement made in Mr. Henshall's memoir of Brindley,[*] to the effect that Brindley's father was the freehold owner of his croft, does not appear to have any foundation ; as the present owner of the property, Dr. Fleming, informs us that it was purchased, about the beginning of the present century, from the heirs of the last of the Heywards, who became its owners in 1688. No such name as Brindley occurs in any of the title-deeds belonging to the property ; and it is probable that the engineer's father was an under-tenant, and merely rented the old cottage in which our hero was born. There is no record of his birth, nor does the name of Brindley appear in the register of the parish of Wormhill, in which the cottage was situated ; but registers in those days were very imperfectly kept, and part of that of Wormhill has been lost.

It is probable that Brindley's father maintained his family by the cultivation of his little croft, and that he was not much, if at all, above the rank of a cottier. It is indeed recorded of him that he was by no means a steady man, and was fonder of sport than of work. He went shooting and hunting, when he should have been labouring ; and if there

* Kippis's 'Biographia Britannica,' Art. Brindley.

was a bull-running within twenty miles, he was
sure to be there. The Bull Ring of the district lay
less than three miles off, at the north end of Long
Ridge Lane, which passed almost by his door; and
of that place of popular resort Brindley's father was
a regular frequenter. These associations led him
into bad company, and very soon reduced him
to poverty. He neglected his children, not only
setting before them a bad example, but permitting
them to grow up without education. Fortunately,
Brindley's mother in a great measure supplied the
father's shortcomings; she did what she could to
teach them what she knew, though that was not
much; but, perhaps more important still, she encouraged them in the formation of good habits by
her own steady industry.*

The different members of the family, of whom
James was the eldest, were thus under the necessity of going out to work at a very early age to
provide for the family wants. James worked at
any ordinary labourer's employment which offered
until he was about seventeen years old. His
mechanical bias had, however, early displayed
itself, and he was especially clever with his knife,
making models of mills, which he set to work in
little mill-streams of his contrivance. It is said
that one of the things in which he took most delight
when a boy, was to visit a neighbouring grist-mill

* Brindley's father seems afterwards to have somewhat recovered himself; for we find him, in 1729, purchasing an undivided
share of a small estate at Lowe Hill, within a mile of Leek, in
Staffordshire, where he had before gone to settle; and he contrived to realise the remaining portion before his death, and
to leave it to his son James. None of the Brindley family
remained at Wormhill, and the name has disappeared in the
district.

and examine the water-wheels, cog-wheels, drum-wheels, and other attached machinery, until he could carry away the details in his head; afterwards imitating the arrangements by means of his knife and such little bits of wood as he could obtain for the purpose. We can thus readily understand how he should have turned his thoughts in the direction in which we afterwards find him employed, and that, encouraged by his mother, he should have determined to bind himself, on the first opportunity that offered, to the business of a millwright.

The demands of trade were so small at the time, that Brindley had no great choice of masters; but at the village of Sutton, near Macclesfield, there lived one Abraham Bennett, a wheelwright and millwright, to whom young Brindley offered himself as apprentice; and in the year 1733, after a few weeks' trial, he became bound to that master for the term of seven years. Although the employment of millwrights was then of a very limited character, they obtained a great deal of valuable practical information whilst carrying on their business. The millwrights were as yet the only engineers. In the course of their trade they worked at the foot-lathe, the carpenter's bench, and the anvil, by turns; thus cultivating the faculties of observation and comparison, acquiring practical knowledge of the strength and qualities of materials, and dexterity in the handling of tools of many different kinds. In country places, where division of labour could not be carried so far as in the larger towns, the mill-wright was compelled to draw largely upon his own resources, and to devise expedients to meet pressing emergencies as they arose. Necessity thus made them dexterous, expert, and skilful in mechanical arrangements, more particularly those connected with mill-work, steam-engines, pumps, cranes, and such like. Hence millwrights in those early days were looked upon as a very important class of workmen. The nature of their business tended to render them self-reliant, and they prided themselves on the importance of their calling. On occasions of difficulty the millwright was invariably resorted to for help ; and as the demand for mechanical skill arose, in course of the progress of manufacturing and agricultural industry, the men trained in millwrights' shops, such as Brindley, Meikle, Rennie, and Fairbairn, were borne up by the force of their practical skill and constructive genius into the highest rank of skilled and scientific engineering.

Brindley, however, only acquired his skill by slow degrees. Indeed, his master thought him slower than most lads, and even stupid. Bennett, like many well-paid master mechanics at that time, was of intemperate habits, and gave very little attention to his apprentice, leaving him to the tender mercies of his journeymen, who were for the most part a rough and drunken set. Much of the lad's time was occupied in running for beer, and when he sought for information he was often met with a rebuff. Skilled workmen were then very jealous of new hands, and those who were in any lucrative employment usually put their shoulders together to exclude outsiders. Brindley had thus to find out nearly everything for himself, and he only worked his way to dexterity through a succession of blunders.

He was frequently left in sole charge of the

wheelwright's shop—the men being absent at jobs
in the country, and the master at the public-house,
from which he could not easily be drawn. Hence,
when customers called at the shop to get any
urgent repairs done, the apprentice was under the
necessity of doing them in the best way he could,
and that often very badly. When the men came
home and found tools blunted and timber spoiled,
they abused Brindley and complained to the master
of his bungling apprentice's handiwork, declaring
him to be a mere " spoiler of wood." On one occasion, when Bennett and the journeymen were
absent, he had to fit in the spokes of a cart-wheel,
and was so intent on completing his job that he did
not find out that he had fitted them all in the wrong
way until he had applied the gauge-stick. Not
long after this occurrence, Brindley was left by
himself in the shop for an entire week, working at
a piece of common enough wheelwright's work,
without any directions ; and he made such a " mess "
of it, that on the master's return he was so enraged,
that he threatened, there and then, to cancel the
indentures and send the young man back to farm-
labourer's work, which Bennett declared was the
only thing for which he was fit.

Brindley had now been two years at the business, and in his master's opinion had learnt next
to nothing ; though it shortly turned out that, notwithstanding the apprentice's many blunders, he
had really groped his way to much valuable practical information on matters relating to his trade.
Bennett's shop would have been a bad school for
an ordinary youth, but it proved a profitable one
for Brindley, who was anxious to learn, and determined to make a way for himself if he could not

I.   M

find one. He must have had a brave spirit to
withstand the many difficulties he had to contend
against, to learn dexterity through blunders, and
success through defeats. But this is necessarily
the case with all self-taught workmen ; and Brindley was mainly self-taught, as we have seen, even
in the details of the business to which he had
bound himself apprentice.

In the autumn of 1735 a small silk-mill at
Macclesfield, the property of Mr. Michael Daintry,
sustained considerable injury from a fire at one
of the gudgeons inside the mill, and Bennett was
called upon to execute the necessary repairs.
Whilst the men were employed at the shop in
executing the new work, Brindley was sent to the
mill to remove the damaged machinery, under the
directions of Mr. James Milner, the superintendent
of the factory. Milner had thus frequent occasion
to enter into conversation with the young man,
and was struck with the pertinence of his remarks
as to the causes of the recent fire and the best
means of avoiding similar accidents in future. He
even applied to Bennett, his master, to permit the
apprentice to assist in executing the repairs of
certain parts of the work, which was reluctantly
assented to. Bennett closely watched his " bungling apprentice," as he called him ; but Brindley,
encouraged by the superintendent of the mill,
succeeded in satisfactorily executing his allotted
portion of the repairs, not less to the surprise of
his master than to the mortification of his men.
Many years after, Brindley, in describing this first
successful piece of mill-work which he had executed, observed, " I can yet remember the delight
which I felt when my work was fixed and fitted

complete; though I could not understand why my master and the other workmen, instead of being pleased, seemed to be dissatisfied with the insertion of every fresh part in its proper place."

The completion of the job was followed by the usual supper and drink at the only tavern in the town, then on Parsonage Green. Brindley's share in the work was a good deal ridiculed by the men when the drink began to operate; on which Mr. Milner, to whose intercession his participation in the work had been entirely attributable, interposed and said, "I will wager a gallon of the best ale in the house, that before the lad's apprenticeship is out he will be a cleverer workman than any here, whether master or man." We have not been informed whether the wager was accepted; but it was long remembered, and Brindley was so often taunted with it by the workmen, that he was not himself allowed to forget that it had been offered. Indeed, from that time forward, he zealously endeavoured so to apply himself as to justify the prediction, for it was nothing less, of his kind friend Mr. Milner; and before the end of his third year's apprenticeship his master was himself constrained to admit that Brindley was not the "fool" and the "blundering blockhead" which he and his men had so often called him.

Very much to the chagrin of the latter, and to the surprise of Bennett himself, the neighbouring millers, when sending for a workman to execute repairs in their machinery, would specially request that "the young man Brindley" should be sent them in preference to any other of the workmen. Some of them would even have the apprentice in preference to the master himself. At this Bennett

M 2

was greatly surprised, and, quite unable to understand the mystery, he even went so far as to inquire
of Brindley where he had obtained his knowledge
of mill-work! Brindley could not tell; it "came
natural-like;" but the whole secret consisted in
Brindley working with his head as well as with
his hands. The apprentice had already been found
peculiarly expert in executing mill repairs, in the
course of which he would frequently suggest
alterations and improvements, more especially in
the application of the water-power, which no one
had before thought of, but which proved to be
founded on correct principles, and worked to the
millers' entire satisfaction. Bennett, on afterwards
inspecting the gearing of one of the mills repaired
by Brindley, found it so securely and substantially
fitted, that he even complained to him of his style
of work. "Jem," said he, "if thou goes on i' this
foolish way o' workin', there will be very little
trade left to be done when thou comes oot o' thy
time: thou knaws firmness o' wark's th' ruin o'
trade." Brindley, however, gave no heed whatever
to the unprincipled suggestion, and considered it
the duty and the pride of the mechanic always to
execute the best possible work.

Among the other jobs which Brindley's master
was employed to execute about this time, was the
machinery of a new paper-mill proposed to be
erected on the river Dane. The arrangements were
to be the same as those adopted in the Smedley
paper-mill on the Irk, and at Throstle-Nest, on
the Irwell, near Manchester; and Bennett went
over to inspect the machinery at those places. But
Brindley was afterwards of opinion that he must
have inspected the taverns in Manchester much

more closely than the paper-mills in the neigh-bourhood; for when he returned, the practical information he brought with him proved almost a blank. Nevertheless, Bennett could not let slip the opportunity of undertaking so lucrative a piece of employment in his special line, and, ill-informed though he was, he set his men to work upon the machinery of the proposed paper-mill.

It very soon appeared that Bennett was alto-gether unfitted for the performance of the contract which he had undertaken. The machinery, when made, would not fit; it would not work; and, what with drink and what with perplexity, Bennett soon got completely bewildered. Yet to give up the job altogether would be to admit his own in-competency as a mechanic, and must necessarily affect his future employment as a millwright. He and his men, therefore, continued distractedly to persevere in their operations, but without the slightest appearance of satisfactory progress.

About this time an old hand, who happened to be passing the place at which the men were at work, looked in upon them and examined what they were about, as a mere matter of curiosity. When he had done so, he went on to the nearest public-house and uttered his sentiments on the subject very freely. He declared that the job was a farce, and that Abraham Bennett was only throw-ing his employer's money away. The statement of what the "experienced hand" had said, was re-peated until it came to the ears of young Brindley. Concerned for the honour of his shop as well as for the credit of his master—though he probably owed him no great obligation on the score either of treatment or instruction—Brindley formed the

immediate resolution of attempting to master the
difficulty so that the work might be brought to a
satisfactory completion.

At the end of the week's work Brindley left the
mill without saying a word of his intention to any
one, and instead of returning to his master's house,
where he lodged, he took the road for Manchester.
Bennett was in a state of great alarm lest he should
have run away; for Brindley, now in the fourth
year of his apprenticeship, had reached the age
of twenty-one, and the master feared that, taking
advantage of his legal majority, he had left his
service never to return. A messenger was de-
spatched in the course of the evening to his
mother's house ; but he was not there. Sunday
came and passed—still no word of young Brindley :
he must have run away!

On Monday morning Bennett went to the paper-
mill to proceed with his fruitless work; and lo! the
first person he saw was Brindley, with his coat off,
working away with greater energy than ever. His
disappearance was soon explained. He had been
to Smedley Mill to inspect the machinery there
with his own eyes, and clear up his master's difficulty. He had walked the twenty-five miles thither
on the Saturday night, and on the following Sunday
morning he had waited on Mr. Appleton, the proprietor of the mill, and requested permission to
inspect the machinery. With an unusual degree of
liberality Mr. Appleton gave the required consent,
and Brindley spent the whole of that Sunday in
the most minute inspection of the entire arrangements of the mill. He could not make notes, but he
stored up the particulars carefully in his head ; and
believing that he had now thoroughly mastered the

difficulty, he set out upon his return journey, and walked the twenty-five miles back to Macclesfield again.

Having given this proof of his determination, as he had already given of his skill in mechanics, Bennett was only too glad to give up the whole conduct of the contract thenceforth to his apprentice; Brindley assuring him that he should now have no difficulty in completing it to his satisfaction. No time was lost in revising the whole design; many parts of the work already fixed were rejected by Brindley, and removed; others, after his own design, were substituted; several entirely new improvements were added; and in the course of a few weeks the work was brought to a conclusion, within the stipulated time, to the satisfaction of the proprietors of the mill.

There was now no longer any question as to the extraordinary mechanical skill of Bennett's apprentice. The old man felt that he had been in a measure saved by young Brindley, and thenceforth, during the remainder of his apprenticeship, he left him in principal charge of the shop. For several years after, Brindley maintained his old master and his family in respectability and comfort; and when Bennett died, Brindley carried on the concern until the work in hand had been completed and the accounts wound up; after which he removed from Macclesfield to begin business on his own account at the town of Leek, in Staffordshire.

CHAPTER II ·

BRINDLEY AS MASTER WHEELWRIGHT AND MILLWRIGHT.

BRINDLEY had now been nine years at his trade, seven as apprentice and two as journeyman; and he began business as a wheelwright at Leek at the age of twenty-six. He had no capital except his skill, and no influence except that which his character as a steady workman gave him. Leek was not a manufacturing place at the time when Brindley began business there in 1742. It was but a small market town, the only mills in the neighbourhood being a few grist-mills driven by the streamlets flowing into the waters of the Dane, the Churnet, and the Trent. These mills usually contained no more than a single pair of stones, and they were comparatively rude and primitive in their arrangement and construction.

Brindley at first obtained but a moderate share of employment. His work was more strongly done, and his charges were consequently higher, than was customary in the district; and the agricultural classes were as yet too poor to enable them to pay the prices of the best work. He gradually, however, acquired a position, and became known for his skill in improving old machinery or inventing such new mechanical arrangements as might be required for any special purpose. He was very careful to execute the jobs which were entrusted to him within the stipulated time, and he began to be spoken of as a thoroughly reliable workman. Thus his business gradually extended to other places at a distance from Leek, and more especially into the Staffordshire Pottery districts, about to rise into importance under the fostering energy of Josiah Wedgwood.

At first Brindley kept neither apprentices nor journeymen, but felled his own timber and cut it up himself, with such assistance as he could procure on the spot. As his business increased he took in an apprentice, and then a journeyman, to carry on the work in the shop while he was absent; and he was often called to a considerable distance from home, more particularly for the purpose of being consulted about any new machinery that was proposed to be put up. Nor did he confine himself to mill-work. He was ready to undertake all sorts of machinery connected with the pumping of water, the draining of mines, the smelting of iron and copper, and the various mechanical arrangements connected with the manufactures rising into importance in the adjoining counties of Cheshire and Lancashire. Whenever he was called upon in this way, he endeavoured to introduce improvements; and to such an extent did he carry this tendency, that he became generally known in the neighbourhood by the name of " The Schemer."

A number of Brindley's memoranda books * are

---

* In the possession of Joseph Mayer, Esq., of Liverpool, who has kindly permitted the author to inspect the whole of his valuable manuscripts relating to Brindley, so curiously illustrative of his start in life as a working and consulting Engineer.

still in existence, which show the varied nature of
his employment during this early part of his career.
It appears from the entries made in them, that he
was not only employed in repairing and fitting up
silk-throwing mills at Macclesfield, all of which
were then driven by water, but also in repairing
corn - mills at Congleton, Newcastle - under - Lyne,
and various other places, besides those in the immediate neighbourhood of Leek, where he lived.
We believe the pocket memoranda books, to which
we refer, were the only records which Brindley
kept of his early business transactions; the rest he
carried in his memory, which by practice became
remarkably retentive. Whilst working as an apprentice at Macclesfield, he had taught himself the
art of writing; but he never mastered it thoroughly,
and to the end of his life he wrote with difficulty,
and almost illegibly. His spelling was also very
bad; and what with the bad spelling and what with
the hieroglyphics in which he wrote, it is sometimes very difficult to decypher the entries made
by him from time to time in his books.

We find him frequently at Trentham. On one
occasion he makes entry of a " Loog of Daal 20 foot
long;" at another time he is fitting a pump for
" Arle Gower," the Earl being one of Brindley's
first patrons. The log of deal, it afterwards appears, was required for the flint-mill of a Mr. Tibots
—"a mow [new?] invontion," as Brindley enters it
in his book—of which more hereafter. On May 18,
1755, he enters " Big Tree to cut 1 day," and he
seems to have felled the tree, and, some months
after, to have cut it up himself, entering so many
days at two shillings a day for the labour. When
he had to travel some distance, he set down

sixpence a day extra for expenses. Thus on one occasion he makes this entry : " For Mr. Kent corn mill of Codan looking out a shaft neer Broun Edge 1 day 0 : 2 : 6."

Between Leek and Trentham lay the then small pottery village of Burslem, which Brindley had frequent occasion to pass through in going to and from his jobs for the Earl. The earthenware then manufactured at Burslem was of a very inferior sort, consisting almost entirely of brown vessels ; and the quantity turned out was so small that it was hawked about on the backs of the potters themselves, or sold by higglers, who carried it from village to village in the panniers of their donkeys.

The Potteries District.

The brothers Elers, the Dutchmen, erected a potwork of an improved kind near Burslem, at the beginning of the century, in which they first practised the art of salt-glazing, brought by them from Holland.

The next improvement introduced was the use of powder of flints, used at first as a wash or dip, and afterwards mixed with tobacco-pipe clay, from which an improved ware was made, called "Flint potters." The merit of introducing this article is usually attributed to William Astbury, of Shelton, who, when on a journey to London, stopping at an inn at Dunstable, noticed the very soft and delicate nature of some burnt flint-stones when mixed with water (the hostler having used the powdered flint as a remedy for a disorder in his horses' eyes), and from thence he is said to have conceived the idea of applying it to the purposes of his trade. In first using the calcined flints, Mr. Astbury's practice was to have them pounded in an iron mortar until perfectly levigated; and being but sparingly used, this answered the demand for some time. But when the use of flint became more common, this tedious process would no longer suffice.

The brothers John and Thomas Wedgwood carried on the pottery business in a very small way, but were nevertheless hampered by an insufficient supply of flint powder, and it was found necessary to adopt some means of increasing it. In their emergency the potters called "The Schemer" to their aid; and hence we find him frequently occupied in erecting flint-mills, in Burslem and the neighbourhood, from that time forward. The success which attended his efforts brought Brindley not only fame, but business.

It happened that, while thus occupied, Mr. John Edensor Heathcote, owner of the Clifton estate near Manchester, became married to one of the daughters of Sir Nigel Gresley, of Knypersley, in the neighbourhood of Burslem, and that the marriage festivities were in progress, when the remarkable ingenuity of the young millwright of Leek was accidentally mentioned in the hearing of Mr. Heathcote one day at dinner. The Manchester man, in the midst of pleasure, did not forget business ; and it occurred to him that this ingenious mechanic might be of use in contriving some method for clearing his Clifton coal-mines of the water by which they had so long been drowned. The old methods of the gin-wheel and tub, and the chain-pump, had been tried, but entirely failed to keep the water under : if this Brindley could but do anything to help him in his difficulty, he would employ him at once ; at all events, he would like to see the man.

Brindley was accordingly sent for, and the whole case was laid before him. Mr. Heathcote described as minutely as possible the nature of the locality, the direction in which the strata lay, and exhibited a plan of the working of the mines. Brindley was perfectly silent for a long time, seemingly absorbed in a consideration of the difficulties to be overcome; but at length his countenance brightened, his eyes sparkled, and he briefly pointed out a method by which he thought he should be enabled, at no great expense, effectually to remedy the evil. His explanations were considered so satisfactory, that he was at once directed to proceed to Clifton with full powers to carry out his proposed plan of operations. This was, to call to his aid the fall of the river Irwell, which formed one boundary of the estate, and pump out the water from the pits by means of the greater power of the water in the river.

With this object Brindley contrived and executed his first tunnel, which he drove through the solid

rock for a distance of six hundred yards, and in this
tunnel he led the river on to the breast of an immense water-wheel fixed in a chamber some thirty
feet below the surface of the ground, from the
lower end of which the water, after exercising its
power, flowed away into the lower level of the
Irwell. The expedient, though bold, was simple,
and it proved effective. The machinery was found
fully equal to the emergency ; and in a very short
time Brindley's wheel and pumps, working night
and day, so cleared the mine of water as to enable
the men to get the coal in places from which they
had long been completely " drowned out."

We are not informed of the remuneration which
the engineer received for carrying out this important work ; but from the entries in his memorandum
book it is probable that all he obtained was only his
workman's wage of two shillings a day. Notwithstanding his ingenuity and hard-working energy,
Brindley never seems, during the early part of his
career, to have earned more than about one-third
the wage of skilled mechanics in our own time ;
and from the insignificant sums charged by him for
expenses, it is clear that he was satisfied to live in
the fashion of an ordinary labourer. What modern
engineers will receive ten guineas a day for doing,
he, with his strong original mind, was quite content
to do for two shillings. But eminent constructive
skill seems to have been lightly appreciated in
those days, if we may judge by the money value
attached to it.*   To this, however, it must be added,

* The low remuneration paid for skilled labour of the same
sort long before Brindley's time is worthy of passing notice. In
1544 John of Padua was paid only two shillings a-day as " devizour
of His Majestie's works," in other words as Royal architect. Still

that at the time of which we speak, the people of
the country were comparatively poor—manufac-
turers as well as landowners. In Macclesfield and the neighbourhood, where
the inventions of men such as Brindley have issued
in so extraordinary a development of wealth, the
operations of trade were as yet in their infancy,
and had numerous obstructions and difficulties to
contend against. Perhaps the greatest difficulty of
all was the absence of those facilities for transport
between one district and another, without which
the existence of trade is simply impossible; but we
shall shortly find Brindley also entering upon this
great work of opening up the internal communications of the country, with an extraordinary degree
of ability and success.

By the middle of last century, Macclesfield and
the neighbouring towns were gradually rising out
of the small button-trade, and aiming at greater
things in the way of manufacture.   In 1755 Mr. N.
Pattison of London, Mr. John Clayton, and a few
other gentlemen, entered into a partnership to

later, Inigo Jones was paid only eight shillings and fourpence a-
day as architect and surveyor of the Whitehall Banqueting House,
and forty-six pounds a-year for house-rent, clerks, and incidental
expenses ; whilst Nicholas Stowe, the master mason, was allowed
but four and tenpence a-day.   When the Duchess of Marlborough
was afterwards engaged in resisting the claims of one of her Blen-
heim surveyors, she told him indignantly "that Sir Christopher
Wren, while employed upon Saint Paul's, was content to be dragged
up to the top of the building three times a-week in a basket, at the
great hazard of his life, for only 200l. a-year"—the actual amount
of his salary as architect of that magnificent Cathedral.   Brindley,
however, fared worse still, and for a long time does not seem to
have risen above mere mechanic's pay, even whilst engaged in
constructing the celebrated canal for the Duke of Bridgewater,
which laid the foundation of so many gigantic fortunes.

build a new silk-mill at Congleton, in Cheshire, on
a larger scale than had yet been attempted in that
neighbourhood. Brindley was employed to execute
the water-wheel and the commoner sort of mill-work
about the building ; but the smaller wheels and the
more complex parts of the machinery, with which
it was not supposed Brindley could be acquainted,
were entrusted to a master joiner and millwright,
named Johnson, who also superintended the progress of the whole work.

The superintendent required Brindley to work
after his mere verbal directions, without the aid of
any plan ; and Brindley was not even allowed to
inspect the models of the machinery required for
the proposed mill. He thus worked at a great dis-
advantage, and the operations connected with the
construction of the intended machinery were very
shortly found in a state of complete muddle. The
proprietors had reason to suspect that their superintendent was not equal to the enterprise which he
had undertaken. At first he endeavoured to assure
them that all was going right ; but at last, after
various efforts, he was obliged to confess his in-
competency and his inability to complete the work.

The proprietors, becoming alarmed, then sent for
Brindley and told him of their dilemma. "Would
he undertake to complete the works ? " He asked
to see the model and plans which the superintendent engineer had proposed to follow out. But on
being applied to, the latter positively refused to
submit his designs to a common millwright, as he
alleged Brindley to be. The proprietors were
almost in despair, and their only reliance now
was on Brindley's genius. " Tell me," he said,
" what is the precise operation that you wish to

perform, and I will endeavour to provide you with the requisite machinery for doing it ; but you must let me carry out the work in my own way." To this they were only too glad to assent ; and having been furnished with the necessary powers, he forthwith set to work.

His intelligent observation of the process of manufacture in the various mills he had inspected, his intimate practical knowledge of machinery of all kinds then in use, and his fertility of resources in matters of mechanical arrangement, enabled him to perform even more than he had promised ; and he not only finished the mill to the complete satisfaction of its owners, but added a number of new and skilful improvements in detail, which afterwards proved of the greatest value. For instance, he adapted lifts to each set of rollers and swifts, by means of which the silk could be wound upon the bobbins equably, instead of in wreaths as in other mills ; and he so arranged the shafting as to throw out of gear and stop either the whole or any part of the machinery at will—an arrangement subsequently adopted in the throstle of the cotton-spinning machine, and, though common enough now, then thought perfectly marvellous. And, in order that the tooth-and-pinion wheels should fit with perfect precision, he expressly invented machinery for their manufacture—a thing that had not before been attempted—all such wheels having, until then, been cut by hand, at great labour and cost. By means of this new machinery, as much work, and of a far better description, could be cut in a day as had before occupied at least a fortnight. The result was, that the new silk-mill, when finished, was found to be one of the most

I.   N

complete and economical arrangements of manufacturing machinery that had up to that time been
erected in the neighbourhood.

After the Congleton silk-mill had been completed, we find Brindley engaged in erecting flint-
mills in the Potteries, of a more powerful and
complete kind than any that had before been tried,
but which were rendered necessary by the growing demands of the earthenware-manufacture. One
of the largest was that erected for Mr. Thomas
Baddely, at a place called Machins' of the Mill, near
Tunstall. We find these entries in Brindley's
pocket-book :—" March 15, 1757. With Mr. Badley
to Matherso about a now flint mill upon a windey
day 1 day 3s. 6d. March 19 draing a plann 1 day
2s. 6d. March 23 draing a plann and to sat out the
wheel race 1 day 4s."

This new mill was driven by water-power, and
the wheel both worked the pumping apparatus
by which the adjoining coal-mine was drained, and
the stamping machinery for pounding and grinding
the flints. The wheel, which was of considerable
diameter, was fixed in a chamber below the surface
of the ground, and the water was conveyed to it
from the mill-pool through a small trough opening upon it at its breast, which kept the paddle-
boxes of the descending part constantly filled,
without any waste whatever, and thus, by the rota-
tion of the wheel, the pumps and stampers were
effectually worked. The main shaft was more
than two hundred yards from the mill ; and to
work the pumps Brindley invented the slide rods,
which were moved horizontally by a crank at the
mill, and gave power to the upright arm of a
crank-lever, whose axis was at the angle, and the

lift at the other extremity. In course of time, as improvements were introduced in the grinding of flints, the stamping apparatus was detached from the machinery; but this water-wheel continued its constant and useful operation of pumping out the mines for full forty years after the death of its inventor; and when it was at length broken up, about the year 1812, the pump-trees, which consisted of wooden staves firmly bound together with ashen hoops, were found to be lined with cow-hides, the working buckets being also covered with leather—a contrivance of which the like, it is believed, has not before been recorded. *

About the same time Brindley was requested by Mr. John Wedgwood to erect a windmill for a similar purpose on an elevated site adjoining the town of Burslem, called The Jenkins; this being one of the first, if not the very first, experiments made of the plan of grinding the calcined flints in water, which in this case was pumped by the action of the machinery from a well situated within the mill itself. This invention, which was of considerable importance, has by some been attributed to Brindley, whose ingenious mind was ever ready to suggest improvements in whatever process of manufacture came under his notice. It was natural that he should closely watch the operation of flint-grinding, having to construct and repair the greater part of the machinery used in the process; and he could not fail to notice the distressing consequences resulting from inhaling the fine particles with which the air of the flint-mills was laden. Hence the probability of his suggesting that the flints should

* 'History of the Borough of Stoke-upon-Trent.' By John Ward, 1853. P. 164.

be ground in water, as calculated not only to prevent waste and preserve the purity of the air, but also to facilitate the operation of grinding,—a simple enough suggestion, but, as the result proved, a most valuable one.

With this object he invented an improved mill, which consisted of a large circular vat, about thirty inches deep, having a central step fixed in the bottom, to carry the axis of a vertical shaft. The moving power was applied to this shaft by a crown cog-wheel placed on the top. At the lower part of the shaft, at right angles to it, were four arms, upon which the grinding-stones were fixed, large blocks of stone of the same kind being likewise placed in the vat. These stones were a very hard silicious mineral, called " Chert," found in abundance in the neighbourhood of Bakewell, in Derbyshire. The broken flints being introduced to the vat and completely covered with water, the axis was made to revolve with great velocity, when the calcined flints were quickly reduced to an impalpable powder. This contrivance of Brindley's proved of great value to Wedgwood, and it was shortly after adopted throughout the Potteries, and continues in use to this day.

Being thus extensively occupied in the invention and erection of machinery driven by one power or another, it was natural that Brindley's attention should have been attracted to the use of steam power in manufacturing operations. Wind and water had heretofore been almost the exclusive agents employed for the purpose; but far-seeing philosophers and ingenious mechanics had for centuries been feeling their way towards the far greater power derived from the pent-up force of vaporised water; and engines had actually been contrived which rendered it likely that the problem would ere long be solved, and a motive agent invented, which should be easily controllable, and independent alike of wind, tides, and waterfalls. Reserving for another place the history of the successive stages of this great invention, it will be sufficient for our present purpose merely to indicate, briefly, the direction of Brindley's labours in this important field.

It appears that Newcomen had as early as the year 1711 erected an atmospheric engine for the purpose of drawing water from a coal mine in the neighbourhood of Wolverhampton; and after considerable difficulties had been experienced in its construction and working, the engine was at length pronounced the most effective and economical that had yet been tried. Other engines of a similar kind were shortly after erected in the coal districts of the north of England, in the tin and copper mines of Cornwall, and in the lead mines of Cumberland, for the purpose of pumping water from the pits.

Brindley, like other contrivers of power, felt curious about this new invention, and proceeded to Wolverhampton to study one of Newcomen's engines erected there. He was greatly struck by its appearance, and, with the irrepressible instinct of the inventor, immediately set about contriving how it might be improved. He found the consumption of coal so great as to preclude its use excepting where coal was unusually abundant and cheap, as, for instance, at the mouth of a coal-pit, where the fuel it consumed was the produce and often the refuse of the mine itself; and he formed the opinion that unless the consumption of coal could be reduced, the extended use of the steam-engine was not practicable, by reason of its dearness, as compared with the power of horses, wind, or water.

With this idea in his head, he proceeded to contrive an improved engine, the main object of which was to ensure greater economy in fuel. In 1756 we find him erecting a steam-engine for one Mr. Broade, at Fenton Vivian, in Staffordshire, in which he adopted the expedient, afterwards tried by James Watt, of wooden cylinders made in the manner of coopers' ware, instead of cylinders of iron. He also substituted wood for iron in the chains which worked at the end of the beam. Like Watt, however, he was under the necessity of abandoning the wooden cylinders; but he surrounded his metal cylinders with a wooden case, filling the intermediate space with wood-ashes; and by this means, and using no more injection of cold water than was necessary for the purpose of condensation, he succeeded in reducing the waste of steam by almost one-half.

Whilst busy with Mr. Broade's engine, we find from the entries in his pocket-book that Brindley occasionally spent several days together at Coal-brookdale, in superintending the making of the boiler-plates, the pipes, and other iron-work. Returning to Fenton Vivian, he proceeded with the erection of his engine-house and the fitting of the machinery, whilst, during five days more, he appears to have been occupied in making the hoops for the cylinders. It takes him five days to get the " great leavor fixed," thirty-nine days to put the boiler together, and thirteen days to get the pit prepared ;

and as he charges only workmen's wages for those days, we infer that the greater part of the work was done by his own hands. He even seems to have himself felled the requisite timber for the work, as we infer from the entry in his pocket-book of "falling big tree 3½ days."

The engine was at length ready after about a year's work, and was set a-going in November, 1757, after which we find these significant entries : "Bad louk [luck] five days ;" then, again, "Bad louk" for three days more ; and, after that, "Midlin louk ;" and so on with "Midlin louk" until the entries under that head come to an end. In the spring of the following year we find him again striving to get his "engon at woork," and it seems at length to have been fairly started on the 19th of March, when we have the entry "Engon at woork 3 days." There is then a stoppage of four days, and again the engine works for seven days more, with a sort of "loud cheer" in the words added to the entry, of "driv a-Heyd !" Other intervals occur, until, on the 16th of April, we have the words "at woor good ordor 3 days," when the entries come to a sudden close.

The engine must certainly have given Brindley a great deal of trouble, and almost driven him to despair, as we now know how very imperfect an engine with wooden hooped cylinders must have been ; and we are not therefore surprised at the entry which he honestly makes in his pocket-book on the 21st of April, immediately after the one last mentioned, when the engine had, doubtless, a second time broken down, "to Run about a Drinking, 0 : 1 : 6." Perhaps he intended the entry to stand there as a warning against giving way to

future despair ; for he underlined the words, as if
to mark them with unusual emphasis.*

Brindley did not remain long in this mood, but
set to work upon the contrivance and erection of
another engine upon a new and improved plan.
What his plan was, may be learnt from the specification lodged in the Patent Office, on the 26th
December, 1758, by "James Brindley, of Leek,
in the county of Stafford, Millwright."† In the
arrangement of this new steam-engine he provided
that the boiler should be made of brick or stone
arched over, and the stove over the fire-place of
cast-iron, fixed within the boiler. The feeding-
pipe for the boiler was to be made with a clack,
opening and shutting by a float upon the surface
of the water in the boiler, which would thus be
self-feeding. The great chains for the segments
at the extremity of the beams were of wood ; and
the pumps were also of wooden staves strongly
hooped together.

Brindley, as a millwright, seems to have long

---

* We find the following memorandum in Brindley's pocket-
book, relating to the expense of working the engine in the year
1760 :—

Miss Clare Maria Broad⁸ fire engine at fentan vivian.

First yeer's work and repare night and day		...	...	£164
Do. torn back ...	...	...	...	...	...	025
Due for tᵉ first yeer ...	...	...	...	...	139
Due for the second yeer	...	...	...	...	102

† He describes it as "A Fire-Engine for Drawing Water out
of Mines, or for Draining of Lands, or for Supplying of Cityes,
Townes, or Gardens with Water, or which may be applicable to
many other great and usefull Purposes, in a better and more
effectual Manner than any Engine or Machine that hath hitherto
been made or used for the like Purpose."—'Specifications of
Patents,' No. 730.

retained his early predilection for wood, and to have preferred it to iron wherever its use was practicable. His plans were, however, subjected to modification and improvement from time to time, as experience suggested; and in the course of a few years, brick, stone, and wood were alike discarded in favour of iron; until, in 1763, we find Brindley erecting a steam-engine for the Walker Colliery, at Newcastle, wholly of iron, manufactured at Coalbrookdale, which was pronounced the most "complete and noble piece of ironwork" that had up to that time been produced.* But by this time Brindley's genius had been turned in another direction; the invention of the steam-engine being now safe in the hands of Watt, who was perseveringly occupied in bringing it to completion.

* Stuart's 'Anecdotes of Steam-Engines,' p. 626.

CHAPTER III.

VERY little had as yet been done to open up the
inland navigation of England, beyond dredging
and clearing out in a very imperfect manner the
channels of some of the larger rivers, so as to
admit of the passage of small barges. Several attempts had been made in Lancashire and Cheshire,
as we have already shown, to open up the navigation of the Mersey and the Irwell from Liverpool
to Manchester. There were similar projects for
improving the Weaver from Frodsham, where it
joins the Mersey, to Winford Bridge above Northwich; and the Douglas, from the Ribble to Wigan.
About the same time like schemes were started
in Yorkshire, with the object of opening up the
navigation of the Aire and Calder to Leeds and
Wakefield, and of the Don from Doncaster to near
Sheffield.

One of the Acts passed by Parliament in 1737
is worthy of notice, as the forerunner of the Bridgewater Canal enterprise: we allude to the Act for
making navigable the Worsley Brook to its junction with the river Irwell, near Manchester. A
similar Act was obtained in 1755, for making navigable the Sankey Brook from the Mersey, about

two miles below Warrington, to St. Helens, Gerrard Bridge, and Penny Bridge. In this case the canal was constructed separate from the brook, but alongside of it; and at several points locks were provided to adapt the canal to the level of the lands passed through.

The same year in which application was made to Parliament for powers to construct the Sankey Canal, the Corporation of Liverpool had under their consideration a much larger scheme—no less than a canal to unite the Trent and the Mersey, and thus open a water-communication between the ports of Liverpool and Hull. It was proposed that the line should proceed by Chester, Stafford, Derby, and Nottingham. A survey was made, principally at the instance of Mr. Hardman, a public spirited merchant of Liverpool, and for many years one of its representatives in Parliament. Another survey was shortly after made at the instance of Earl Gower, afterwards Marquis of Stafford, and it was probably in making this survey that Brindley's attention was first directed to the business of canal engineering.

We find his first entry relating to the subject made on the 5th of February, 1758—"novocion [navigation] 5 days;" the second, a little better spelt, on the 19th of the same month—"a bout the novogation 3 days;" and afterwards — "surveing the novogation from Long brigg to Kinges Milles 12 days ½." It does not, however, appear that the scheme made much progress, or that steps were taken at that time to bring the measure before Parliament; and Brindley continued to pursue his other employments, more especially the erection of "fire-engines" after his new patent. This

continued until the following year, when we find
him in close consultation with the Duke of Bridgewater relative to the construction of his proposed
canal from Worsley to Manchester.

The early career of this distinguished nobleman
was of a somewhat remarkable character. He was
born in 1736, the fifth and youngest son of Scroop,
third Earl and first Duke of Bridgewater, by Lady
Rachel Russell. He lost his father when only
five years old, and all his brothers died by the
time that he had reached his twelfth year, at which
early age he succeeded to the title of Duke of
Bridgewater. He was a weak and sickly child,
and his mental capacity was thought so defective,
that steps were even in contemplation to set him
aside in favour of the next heir to the title and
estates. His mother seems almost entirely to have
neglected him. In the first year of her widowhood
she married Sir Richard Lyttleton, and from that
time forward took the least possible notice of her
boy.

The young Duke did not give much promise
of surviving his consumptive brothers, and his
mind was considered so incapable of improvement,
that he was left in a great measure without either
domestic guidance or intellectual discipline and
culture. Horace Walpole writes to Mann in 1761 :
"You will be happy in Sir Richard Lyttleton and
his Duchess ; they are the best-humoured people
in the world." But the good humour of this handsome couple was mostly displayed in the world of
gay life, very little of it being reserved for home
use. Possibly, however, it may have been even
fortunate for the young Duke that he was left so
much to himself, to profit by the wholesome neglect

of special nurses and tutors, who are not always the most judicious in their bringing up of delicate children.

At seventeen, the young Duke's guardians, the Duke of Bedford and Lord Trentham, finding him still alive and likely to live, determined to send him abroad on his travels—the wisest thing they could have done. They selected for his tutor the celebrated traveller, Robert Wood, author of the well-known work on Troy, Baalbec, and Palmyra; afterwards appointed to the office of Under-Secretary of State by the Earl of Chatham. Wood was an accomplished scholar, a persevering traveller, and withal a man of good business qualities. His habits of intelligent observation could not fail to be of service to his pupil, and it is not unnatural to suppose that the great artificial watercourses and canals which they saw in the course of their travels had some effect in afterwards determining the latter to undertake the important works of a similar character by which his name became so famous. While passing through the south of France, the Duke was especially interested by his inspection of the Grand Canal of Languedoc, a magnificent work executed under great difficulties, and which had promoted in an extraordinary degree the prosperity of that part of the kingdom.* Proceeding into Italy, the Duke and his companion inspected all that was worthy of being seen there, including the picture galleries at Florence, Venice, and Rome. During their visit Mr. Wood sat to Mengs for his portrait, which still forms part of the Bridgewater collection. The Duke also purchased

---

* See Appendix, *Grand Canal of Languedoc*, and its execution by Riquet de Bonrepos.

works of sculpture at Rome; but that he himself entertained no great enthusiasm for art is evident from the fact related by the late Earl of Elles- mere, that these works remained in their original packing-cases until after his death.*

Returned to England, he seems to have led the usual life of a gay young nobleman of the time, with plenty of money at his command. In 1756, when only twenty years old, he appears from the 'Racing Calendar' to have kept race-horses; occa- sionally riding them in matches himself. Though in after life a very bulky man, he was so light as a youth, that on one occasion Lord Ellesmere says a bet was jokingly offered that he would be blown off his horse. Dressed in a livery of blue silk and silver, with a jockey cap, he once rode a race against His Royal Highness the Duke of Cumberland, on the long terrace at the back of the wood in Trent- ham Park, the seat of his relative, Earl Gower. During His Royal Highness's visit, the large old green-house, since taken down, was hastily run up for the playing of skittles; and prison bars and other village games were instituted for the recrea- tion of the guests. Those occupations of the Duke were varied by an occasional visit to his racing- stud at Newmarket, where he had a house for some time, and by the usual round of London gaieties during the season.

A young nobleman of tender age, moving freely in circles where were to be seen some of the finest specimens of female beauty in the world, could scarcely be expected to pass heart-whole; and hence the occurrence of the event in his London

---

* 'Essays in History, Biography, Geography, Engineering,' &c. By the late Earl of Ellesmere. London, 1858. P. 226.

life which, singularly enough, is said to have driven
him in a great measure from society, and induced
him to devote himself to the construction of canals!
We find various allusions in the letters of the
time to the intended marriage of the young Duke
of Bridgewater. One rumour pointed to the only
daughter and heiress of Mr. Thomas Revell, formerly M.P. for Dover, as the object of his choice.
But it appears that the lady to whom he became the
most strongly attached was one of the Gunnings—
the comparatively portionless daughters of an Irish
gentleman, who were then the reigning beauties
at Court. The object of the Duke's affection was
Elizabeth, the youngest daughter, and perhaps
the most beautiful of the three. She had been
married to the fourth Duke of Hamilton, in Keith's
Chapel, Mayfair, in 1752, "with a ring of the bed-
curtain, half-an-hour after twelve at night," * but
the Duke dying shortly after, she was now a gay
and beautiful widow, with many lovers in her
train. In the same year in which she had been
clandestinely married to the Duke of Hamilton,
her eldest sister was married to the sixth Earl
of Coventry.

The Duke of Bridgewater paid his court to the
young widow, proposed, and was accepted. The
arrangements for the marriage were in progress,
when certain rumours reached his ear reflecting
upon the character of Lady Coventry, his intended
bride's elder sister, who was certainly more fair
than she was wise. Believing the reports, he required the Duchess to desist from further intimacy
with her sister, a condition which her high spirit
would not brook, and, the Duke remaining firm,

* 'Walpole to Mann,' Feb. 27th, 1752.

the match was broken off. From that time forward he is said never to have addressed another woman in the language of gallantry.*

The Duchess of Hamilton, however, did not remain long a widow. In the course of a few months she was engaged to, and afterwards married, John Campbell, subsequently Duke of Argyll. Horace Walpole, writing of the affair to Marshal Conway, January 28th, 1759, says : "You and M. de Bareil do not exchange prisoners with half as much alacrity as Jack Campbell and the Duchess of Hamilton have exchanged hearts. . . . It is the prettiest match in the world since yours, and everybody likes it but the Duke of Bridgewater and Lord Conway. What an extraordinary fate is attached to these two women! Who could have believed that a Gunning would unite the two great houses of Campbell and Hamilton? For my part, I expect to see my Lady Coventry Queen of Prussia. I would not venture to marry either of them these thirty years, for fear of being shuffled out of the world prematurely to make room for the rest of their adventures."

The Duke, like a wise man, sought consolation for his disappointment by entering into active and useful occupation. Instead of retiring to his beautiful seat at Ashridge, we find him straightway proceeding to his estate at Worsley, on the borders of Chat Moss, in Lancashire, and conferring with John Gilbert, his land-steward, as to the

---

* Chalmers, in his 'Biographical Dictionary,' vol. xiii., 94, gives another account of the rumoured cause of the Duke's subsequent antipathy to women ; but the above statement of the late Earl of Ellesmere, confirmed as it is by certain passages in Walpole's Letters, is more likely to be the correct one.

Portrait of the Young Duke.

I.   O

practicability of cutting a canal by which the coals found upon his Worsley estate might be readily conveyed to market at Manchester. Manchester and Liverpool at that time were improving towns, gradually rising in importance and increasing in population. The former place had long been noted for its manufacture of coarse cottons, or "coatings," made of wool, in imitation of the goods known on the Continent by that name. The Manchester people also made fustians, mixed stuffs, and small wares, amongst which leather-laces for women's bodices, shoe-ties, and points were the more important. But the operations of manufacture were still carried on in a clumsy way, entirely by hand. The wool was spun into yarn by means of the common spinning wheel, for the spinning-jenny had not yet been invented, and the yarn was woven into cloth by the common hand-loom. There was no whirr of engine-wheels then to be heard; for Watt's steam-engine had not yet come into existence. The air was free from smoke, except that which arose from household fires, and there was not a single factory-chimney in Manchester.

In 1724, Dr. Stukeley says Manchester contained no fewer than 2400 families, and that their trade was "incredibly large" in tapes, ticking, girth-webb, and fustians. In 1757 the united population of Manchester and Salford was only 20,000; * it is now, after the lapse of a century, 460,000! The Manchester manufacturer was then a very humble personage compared with his modern representative. He was part chapman, part weaver, and

* Aikin's 'Description of the Country from Thirty to Forty Miles round Manchester.' London, 1795.

part merchant—working hard, living frugally, principally on oatmeal, and usually contriving to save a little money.

Dr. Aikin, writing in 1795, thus described the Manchester manufacturer in the first half of the eighteenth century : "An eminent manufacturer in that age," said he, "used to be in his warehouse before six in the morning, accompanied by his children and apprentices. At seven they all came

View of Manchester in 1740.
[Fac-simile of an Engraving of the period.]

in to breakfast, which consisted of one large dish of water-pottage, made of oatmeal, water, and a little salt, boiled thick, and poured into a dish. At the side was a pan or basin of milk, and the master and apprentices, each with a wooden spoon in his hand, without loss of time, dipped into the same dish, and thence into the milk-pan, and as soon as it was finished they all returned to their work." What a contrast to the "eminent manufacturer" of our own day !

As trade increased, its operations became more

subdivided, and special classes and ranks began to spring into importance. The manufacturers sent out riders to take orders, and gangs of chapmen with pack-horses to distribute the goods and bring back wool, which they either used up themselves, or sold to makers of worsted yarn at Manchester, or to the clothiers of Rochdale, Saddleworth, or the West Riding of Yorkshire. Mr. Walker, author of the 'Original,' left the following interesting reminiscence of the dealings of Manchester men with the inhabitants of the Fen districts :—" I have by tradition," said he, " the following particulars of the mode of carrying on the home trade by one of the principal merchants of Manchester, who was born at the commencement of the last century, and who realised a sufficient fortune to keep a carriage when not half a dozen were kept in the town by persons connected with business. He sent the manufactures of the place into Nottinghamshire, Lincolnshire, Cambridgeshire, and the intervening counties, and principally took in exchange feathers from Lincolnshire, and malt from Cambridgeshire and Nottinghamshire. All his commodities were conveyed on pack-horses, and he was from home the greater part of every year, performing his journeys entirely on horseback. His balances were received in guineas, and were carried with him in his saddle-bags. He was exposed to the vicissitudes of the weather, to great labour and fatigue, and to constant danger. In Lincolnshire he travelled chiefly along bridle-ways through fields where frequent gibbets warned him of his perils, and where flocks of wild fowl continually darkened the air. Business carried on in this manner required a combination of personal attention, courage, and physical strength, not to be hoped for in a deputy ; and a merchant then led a much more severe and irksome life than a bagman afterwards, and still more than a traveller of the present day. In the earlier days of the merchant abovementioned, the wine merchant who supplied Manchester, resided at Preston, then always called Proud Preston, because exclusively inhabited by gentry. The wine was carried on horses, and a gallon was considered a large order. Men in business confined themselves generally to punch and ale, using wine only as a medicine, or on extraordinary occasions; so that a considerable tradesman somewhat injured his credit amongst his neighbours by being so extravagant as to send to a tavern for wine, to entertain a London customer." *

The roads out of Manchester in different directions, like those in most districts throughout the kingdom, were in a very neglected state, being for the most part altogether impracticable for waggons. Hence the use of pack-horses was an absolute necessity ; and the roads were but ill-adapted even for them. Indeed, it was more difficult then to reach a village twenty miles out of Manchester than it is to make the journey from thence to London now. The only coach to London plied but every second day, and it was four days and a half in making the journey, there being a post only three times a week.† The roads in most districts

---

* Thomas Walker: The Original, No. xi. Article entitled " Change in Commerce."

† March 3rd, 1760, the Flying Machine was started, and advertised to perform the journey, "if God permit," in three days, by John Hanforth Matthew Howe, Samuel Granville, and William Richardson. Fare inside, 2*l.* 5*s.* ; outside, half-price.

of Lancashire were what were called "mill roads," along which a horse with a load of oats upon its back might proceed towards the mill where they were to be ground. There was no private carriage kept by any person in business in Manchester until the year 1758, when the first was set up by some specially luxurious individual. But wealth led to increase of expenditure, and Aikin mentions that there was "an evening club of the most opulent manufacturers, at which the expenses of each person were fixed at fourpence-halfpenny— fourpence for ale, and a halfpenny for tobacco." The progress of luxury was further aided by the holding of a dancing assembly once a week in a room situated about the middle of King Street, now a busy thoroughfare, the charge for admission to the nightly ball being half-a-crown the quarter. The ladies had their maids to wait for them with lanterns and pattens, and to conduct them home ; "nor," adds Aikin, "was it unusual for their partners also to attend them."

The imperfect state of the communications lead- ing to and from Manchester rendered it a matter of some difficulty at certain seasons to provide food for so large a population. In winter, when the roads were closed, the place was in the condition of a beleaguered town ; and even in summer, the land about Manchester itself being comparatively sterile, the place was badly supplied with fruit, vegetables, and potatoes, which, being brought from considerable distances slung across horses' backs, were so dear as to be beyond the reach of the mass of the population. The distress caused by this frequent dearth of provisions was not effec- tually remedied until the canal navigation became

completely opened up. Thus a great scarcity
of food occurred in Manchester and the neighbourhood in 1757, which the common people attributed
to the millers and corndealers; and unfortunately
the notion was not confined to the poor who were
starving, but was equally entertained by the well-
to-do classes who had enough to eat. An epigram
by Dr. Byrom, the town clergyman, written in 1737,
on two millers (tenants of the School corn-mills),
who, from their spare habits, had been nick-named
" Skin " and " Bone," was now revived, and tended to
fan the popular fury. It ran thus :—

> "Bone and Skin, two millers thin,
>   Would starve the town, or near it ;
> But be it known to Skin and Bone,
>   That Flesh and Blood can't bear it."

The popular hunger and excitement increasing, at
length broke out in open outrage ; and a riot took
place in 1758, long after remembered in Manchester
as the " Shude Hill fight," in which unhappily
several lives were lost.

For the same reasons, the supply of coals was
but scanty in winter; and though abundance of
the article lay underground, within a few miles of
Manchester, in nearly every direction, those few
miles of transport, in the then state of the roads,
were an almost insurmountable difficulty. The
coals were sold at the pit mouth at so much the
horse-load, weighing 280 lbs., and measuring two
baskets, each thirty inches by twenty, and ten
inches deep ; that is, as much as an average horse
could carry on its back.* The price of the coals at

---

* This "load" is still used as a measure of weight, though the
practice of carrying all sorts of commodities on horses' backs, in
which it originated, has long since ceased.

the pit mouth was 10*d*. the horse-load ; but by the time the article reached the door of the consumer in Manchester, the price was usually more than doubled, in consequence of the difficulty and cost of conveyance. The carriage alone amounted to about nine or ten shillings the ton.

There was as yet no connection of the navigation of the Mersey and Irwell with any of the collieries situated to the eastward of Manchester, by which a supply could reach the town in boats ; and although the Duke's collieries were only a comparatively short distance from the Irwell, the coals had to be carried on horses' backs or in carts from the pits to the river to be loaded, and after reaching Manchester they had again to be carried to the doors of the consumers,—so that there was little if any saving to be effected by that route. Besides, the minimum charge insisted on by the Mersey Navigation Company of 3*s*. 4*d*. a ton for even the shortest distance, proved an effectual barrier against any coal reaching Manchester by the river.

The same difficulty stood in the way of the transit of goods between Manchester and Liverpool. By road the charge was 40*s*. a ton, and by river 12*s*. a ton ; that between Warrington and Manchester being 10*s*. a ton : besides, there was great risk of delay, loss, and damage by the way. Some idea of the tediousness of the river navigation may be formed from the fact, that the boats were dragged up and down stream exclusively by the labour of men, and that horses and mules were not employed for this purpose until after the Duke's canal had been made. It was, indeed, obvious that unless some means could be devised for facilitating and

cheapening the cost of transport between the seaport and the manufacturing towns, there was little
prospect of any considerable further development
being effected in the industry of the district.

Such was the state of things when the Duke of
Bridgewater turned his attention to the making
of a water-road for the passage of his coal from
Worsley to Manchester. The Old Mersey Company would give him no facilities for sending his
coals by their navigation, but levied the full charge
of 3s. 4d. for every ton he might send to Manchester
by river even in his own boats. He therefore
perceived that to obtain a vend for his article, it
was necessary he should make a way for himself;
and it became obvious to him that if he could but
form a canal between the two points, he would at
once be enabled to secure a ready sale for all the
coals that he could raise from his Worsley pits.

We have already stated that, as early as 1737,
an Act had been obtained by the Duke's father,
giving power to make the Worsley Brook navigable
from the neighbourhood of the pits to the Irwell.
But the enterprise, and its cost, appear to have
been too formidable; so the powers of the Act
were allowed to expire without anything being
done to carry them out. The young Duke now
determined to revive the Act in another form, and
in the early part of 1759 he applied to Parliament
for the requisite powers to enable him to cut a
navigable canal from Worsley Mill eastward to
Salford, and to carry the same westward to a
point on the river Mersey, called Hollin Ferry. He
introduced into the bill several important concessions to the inhabitants of Manchester. He bound
himself not to exceed the freight of 2s. 6d. per ton

on all coals brought from Worsley to Manchester, and not to sell the coal so brought from the mines to that town at more than 4*d*. per hundred, which was less than half the then average price. It was clear that, if such a canal could be made and the navigation opened as proposed, it would prove a great public boon to the inhabitants of Manchester. The bill was accordingly well supported, and it passed the legislature without opposition, receiving the Royal assent in March, 1759.

The Duke gave further indications of his promptitude and energy, in the steps which he adopted to have the works carried out without loss of time. He had no intention of allowing the powers of this Act to remain a dead letter, as the former had done. Accordingly, no sooner had it passed than he set out for his seat at Worsley to take the requisite measures for constructing the canal. The Duke was fortunate in having for his land-agent a very shrewd, practical, and enterprising person, in John Gilbert, whom he consulted on all occasions of difficulty.

Mr. Gilbert was the brother of Thomas Gilbert, the originator of the Gilbert Unions, then agent to the Duke's brother-in-law, Lord Gower. That nobleman had for some time been promoting the survey of a canal to unite the Mersey and the Trent, on which Brindley had been employed, and thus became known to Gilbert as well as to his brother. We find from an entry in Brindley's pocket-book that the millwright had sundry interviews with Thomas Gilbert on matters of business previous to the passing of the first Bridgewater Canal Bill, though there is no evidence that he was employed in making the survey. Indeed, it is questionable

whether any survey was made of the first scheme.
Engineering projects were then submitted to Parliamentary Committees in a very rough state.
Levels were guessed at rather than surveyed and
calculated; and merely general powers were taken
enabling such property to be purchased as might

Map of the Duke's Canal.

[Western Part.]

by possibility be required for the execution of
the works. In the case of the Bridgewater Canal,
the prices of land and compensation for damage
were directed to be assessed by a local committee
appointed by the Act for the purpose.

When the Duke proceeded to consider with
Gilbert the best mode of carrying out the proposed

canal, it appeared clear to them that the plan originally contemplated was faulty in many respects,
and that an application must be made to Parliament for further powers. By the original Act it
was intended to descend from the level of the coal-
mines at Worsley by a series of locks into the river

Map of the Duke's Canal.

[Eastern Part.]

Irwell. This, it was found, would necessarily involve a heavy cost both in the construction and
working of the canal, as well as considerable delay
in the conduct of the traffic, which it was most
desirable to avoid. Neither the Duke nor Gilbert
had any practical knowledge of engineering; nor,
indeed, were there many men in the country at

that time who knew much of the subject ; for it
must be remembered that this canal of the Duke's
was the very first project in England for cutting
a navigable trench through the dry land, and
carrying merchandise in it across the country,
independent of the course of the existing streams.

It was in this emergency that Gilbert advised
the Duke to call to his aid James Brindley, whose
fertility of resources and skill in overcoming
mechanical difficulties had long been the theme of
general admiration in his own district. Doubtless
the Duke was as much impressed by the native
vigour and originality of the unlettered genius
introduced to him by his agent, as were all with
whom he was brought in contact. Certain it is
that the Duke showed his confidence in Brindley
by entrusting him with the conduct of the proposed
work ; and, as the first step, he was desired to go
over the ground at once, and give his opinion as
to the best plan to be adopted for carrying it out
with despatch.

Brindley, accordingly, after making what he
termed an "ochilor [ocular] servey or a ricconitor-
ing," speedily formed his conclusion, and came back
to the Duke with his advice. It was that, instead
of carrying the canal down into the Irwell by a
flight of locks, and so up again on the other side
to the proposed level, it should be carried right
over the river, and constructed upon one uniform
level throughout. But this, it was clear, would
involve a series of formidable works, the like of
which had not before been attempted in England.
In the first place, the low ground on the north
side of the Irwell would have to be filled up by a
massive embankment, and to be united with the

land on the other bank by means of a large aqueduct of stone. Would it be practicable or possible to execute works of such magnitude? Brindley expressed so strong and decided an opinion of their practicability, that the Duke was won over to his views, and determined again to go to Parliament for the requisite powers to enable him to carry out the design.

Many were the deliberations which took place about this time between the Duke, Gilbert, and Brindley, in the Old Hall at Worsley, where the Duke had now taken up his abode. We find from Brindley's pocket-book memoranda, that in the month of July, 1759, he had taken up his temporary quarters at the Old Hall; and from time to time, in the course of the same year, while the details of the plan were being prepared with a view to the intended application to Parliament, he occasionally stayed with the Duke for weeks together. He made a detailed survey of the new line, and at the same time, in order to facilitate the completion of the undertaking when the new powers had been obtained, he proceeded with the construction of the sough or level at Worsley Mill, and such other portions of the work as could be executed under the original powers.

During the same period Brindley travelled backwards and forwards a great deal, on matters connected with his various business in the Pottery district. We find, from his private record, that he was occupied at intervals in carrying forward his survey of the proposed canal through Staffordshire, visiting with this object the neighbourhood of Newcastle-under-Lyne, Lichfield, and Tamworth. He also continued to give his attention to mills,

water-wheels, cranes, and fire-engines, which he
had erected, or which required repairs, in various
parts of the same district. In short, he seems at
this period to have been fully employed as a millwright; and although, as we have seen, the remuneration which he received for his skill was
comparatively small, being a man of frugal habits
he had saved a little money; for about this time
we find him able to raise a sum of 543*l.* 6*s.* 8*d.*, being
his fourth share of the purchase-money of the
Turnhurst estate, situated near Golden Hill, in the
county of Stafford.

The principal part of this sum was no doubt
borrowed, as appears by his own memoranda,
from his friend Mr. Launcelot, of Leek; but the
circumstance proves that, amongst his townsmen
and neighbours, who knew him best, he stood in
good credit and repute. His other partners in the
purchase were Mr. Thomas Gilbert (Earl Gower's
agent), Mr. Henshall (afterwards his brother-in-
law), and his brother John Brindley. The estate
was understood to be full of minerals, the knowledge of which had most probably been obtained by
Brindley in the course of his surveying of the proposed Staffordshire canal; and we shall afterwards
find that he turned the purchase to good account.

At length the new plans of the canal from
Worsley to Manchester were completed and ready
for deposit; and on the 23rd of January, after a
visit to the Duke and Gilbert at the Hall, we find
the entry in Brindley's pocket-book of "Sot out
for London." On the occasion of his visits to
London, Brindley adopted the then most convenient method of travelling on horseback, the
journey usually occupying five days. We find

him varying his route according to the state of the weather and of the roads. In summer he was accustomed to go by Coventry, but in winter he made for the Great North Road by Northampton, which was usually in better condition for winter travelling.

The second Act passed like the first, without opposition, early in the session of 1760. It enabled the Duke to carry his proposed canal *over* the river Irwell, near Barton Bridge, some five miles westward of Manchester, by means of a series of arches, and to vary its course accordingly; whilst it further authorised him to extend a short branch to Longford Bridge, near Stretford,—that to Hollin Ferry, authorised by the original Act, being abandoned. In the mean time the works near Worsley had been actively pushed forward, and considerable progress had been made by the time the additional powers had been obtained. That part of the canal which lay between Worsley Mill and the public highway leading from Manchester to Warrington had been cut; the sough or level between Worsley Mill and Middlewood, for the purpose of supplying water to the canal, was considerably advanced; and operations had also been begun in the neighbourhood of Salford and on the south of the river Irwell.

The most difficult part of the undertaking, however, was that authorised by the new Act; and the Duke looked forward to its execution with the greatest possible anxiety. Although aqueducts of a far more formidable description had been executed abroad, nothing of the kind had until then been projected in this country ; and many regarded the plan of Brindley as altogether wild and impracticable. The proposal to confine and carry a body

of water within a water-tight trunk of earth upon
the top of an embankment across the low grounds
on either side of the Irwell, was considered foolish
and impossible enough; but to propose to carry
ships upon a lofty bridge, over the heads of other
ships navigating the Irwell which flowed underneath, was laughed at as the dream of a madman.
Brindley, by leaving the beaten path, thus found
himself exposed to the usual penalties which befall
originality and genius.

The Duke was expostulated with by his friends,
and strongly advised not to throw away his money
upon so desperate an undertaking. Who ever heard
of so large a body of water being carried over
another in the manner proposed? Brindley was
himself appealed to; but he could only repeat his
conviction as to the entire practicability of his
design. At length, by his own desire and to allay
the Duke's apprehensions, another engineer was
called in and consulted as to the scheme. To
Brindley's surprise and dismay, the person consulted concurred in the view so strongly expressed
by the public. He characterised the plan of the
Barton aqueduct and embankment as instinct with
recklessness and folly; and after expressing his
unqualified opinion as to the impracticability of
executing the design, he concluded his report to
the Duke thus: "I have often heard of castles in
the air; but never before saw where any of them
were to be erected." *

* We have heard the name of Smeaton mentioned as that of
the engineer consulted on the occasion, but are unable to speak
with certainty on the point. Excepting Smeaton, however, there
was then no other engineer in the country of recognised eminence
in the profession.

It is to the credit of his Grace that, notwithstanding these strong adverse opinions, he continued to give his confidence to the engineer whom he had selected to carry out the work. Brindley's common-sense explanations, though they might not remove all his doubts, nevertheless determined the Duke to give him the full opportunity of carrying out his design; and he was accordingly authorised to proceed with the erection of his "castle in the air." Its progress was watched with great interest, and people flocked from all parts to see it.

The Barton aqueduct is about two hundred yards in length and twelve yards wide, the centre part being sustained by a bridge of three semicircular arches, the middle one being of sixty-three feet span. It carries the canal over the Irwell at a height of thirty-nine feet above the river—this head-room being sufficient to enable the largest barges to pass underneath without lowering their masts. The bridge is entirely of stone blocks, those on the faces being dressed on the front, beds, and joints, and cramped with iron. The canal, in passing over the arches, is confined within a puddled * channel to prevent leakage,

---

* The process of puddling is of considerable importance in canal engineering. Puddle is formed by a mixture of well-tempered clay and sand reduced to a semi-fluid state, and rendered impervious to water by manual labour, as by working and chopping it about with spades. It is usually applied in three or more strata to a depth or thickness of about three feet : and care is taken at each operation so to work the new layer of puddling stuff as to unite it with the stratum immediately beneath. Over the top course a layer of common soil is usually laid. It is only by the careful employment of puddling that the filtration of the water of canals into the neighbouring lower lands through which they pass can be effectually prevented.

and is in as good a state now as on the day on which it was completed. Although the Barton aqueduct has since been thrown into the shade by the vastly greater works of modern engineers,

Barton Aqueduct.
[By Percival Skelton, after his original Drawing.]

it was unquestionably a very bold and ingenious enterprise, if we take into account the time at which it was erected. Humble though it now appears, it was the parent of the magnificent aqueducts of Rennie and Telford, and of the viaducts of Stephenson and Brunel, which rival the greatest works of any age or country.

The embankments formed across the low grounds on either side of the Barton viaduct were also considered very formidable works at that day. A contemporary writer speaks of the embankment across Stretford Meadows as an amazing bank of earth 900 yards long, 112 feet in breadth across the base, 24 feet at top, and 17 feet high. The greatest difficulty anticipated, was the holding of so large a body of water within a hollow channel formed of soft materials. It was supposed at first that the water would soak through the bank, which its weight would soon burst, and wash away all before it. But Brindley, in the course of his experience, had learnt something of the powers of clay-puddle to resist the passage of water. He had already succeeded in stopping the breaches of rivers flowing through low grounds by this means; and the thorough manner in which he finished the bed of this canal, and made it impervious to water, may be cited as a notable illustration of the engineer's practical skill, taking into account the early period at which this work was executed.

Not the least difficult part of the undertaking was the formation of the canal across Trafford Moss, where the weight of the embankment pressed down and "blew up" the soft oozy stuff on each side; but the difficulty was again overcome by the engineer's specific of clay-puddle, which proved completely successful. Indeed, the execution of these embankments by Brindley was regarded at the time as something quite as

extraordinary in their way as the erection of the
Barton aqueduct itself.

The rest of the canal between Longford and
Manchester, being mostly on sidelong ground, was
cut down on the upper side and embanked up on
the other by means of the excavated earth. This
was comparatively easy work; but a matter of
greater difficulty was to accommodate the streams
which flowed across the course of the canal. This
was, however, provided for in a highly ingenious
manner. For instance, a stream called Cornbrook
was found too high to pass under the canal at its
natural level. Accordingly, Brindley contrived a
weir, over which the stream fell into a large basin,
from whence it flowed into a smaller one open at
the bottom. From this point a culvert, constructed
under the bed of the canal, carried the waters
across to a well situated on its further side,
where the waters rising up to their natural level,
again flowed away in their proper channel. A
similar expedient was adopted at the Manchester
terminus of the canal, at the point at which it
joined the waters of the Medlock.

It was a principle of Brindley's never to permit
the waters of any river or brook to intermix
with those of the canal except for the purpose
of supply; as it was clear that in a time of flood
such intermingling would be a source of great
danger to the navigation. In order, therefore, to
provide for the free passage of the Medlock without causing a rush into the canal, a weir was
contrived 366 yards in circumference, over which
its waters flowed into a lower level, and from
thence into a well several yards in depth, down
which the whole river fell. It was received at

the bottom in a subterranean passage, by which it passed into the river Irwell, near at hand. The weir was very ingeniously contrived, though it was afterwards found necessary to make considerable alterations and improvements in it, as experience suggested, in order effectually to accommodate the flood-waters of the Medlock. Arthur Young, when visiting the canal, shortly after it was opened up to Manchester, says, "The whole plan of these works shows a capacity and extent of mind which foresees difficulties, and invents remedies in anticipation of possible evils. The connection and dependence of the parts upon each other are happily imagined; and all are exerted in concert, to command by every means the wished-for success." *

Brindley's labours, however, were not confined to the construction of the canal, but his attention seems to have been equally directed to the contrivance of the whole of the arrangements and machinery by which it was worked. The open navigation between Worsley Mill and Manchester was 10¼ miles in length. A large basin was excavated at the former place, of sufficient capacity to contain a great many boats, and to serve as a head for the navigation.

It is at Worsley Basin that the canal enters the bottom of the hill by a subterranean channel which extends for a great distance,—connecting the different workings of the mine,—so that the coals can be readily transported in boats to their place of sale. A representation of the basin is given on page 217. It lies at the base of a cliff of sandstone, some hundred feet in height, overhung by

* 'Six Months' Tour through the North of England,' vol. iii., p. 258. Ed. 1770.

luxuriant foliage, beyond which is seen the graceful spire of Worsley church. In contrast to this scenic beauty above, lies the almost stagnant pool beneath. The barges * laden with coal emerge from the mine through the two low, semi-circular arches opening at the base of the rock, such being the entrances to the underground workings. The smaller aperture is the mouth of a canal of only half a mile in length, serving to prevent the obstruction which would be caused by the entrance and egress of so many barges through a single passage. The other archway is the entrance of a wider channel, extending nearly six miles in the direction of Bolton, from which various other canals diverge in different directions.

In Brindley's time, this subterranean canal, hewn out of the rock, was only about a mile in length, but it now extends to nearly forty miles in all directions underground. Where the tunnel passed through earth or coal, the arching was of brickwork ; but where it passed through rock, it was simply hewn out. This tunnel acts not only as a drain and water-feeder for the canal itself, but as a means of carrying the facilities of the navigation through the very heart of the collieries ; and it will readily be seen of how great a value it must have proved in the economical working of the navigation, as well as of the mines, so far as the traffic in coals was concerned.

At every point Brindley's originality and skill

* The barges are narrow and long, each conveying about ten tons of coal. They are drawn along the tunnels by means of staples fastened to the sides. When they are empty, and consequently higher in the water, they are so near the roof that the bargemen, lying on their backs, can propel them with their feet.

were at work. He invented the cranes for the
purpose of more readily loading the boats with the
boxes filled with the Duke's "black diamonds."

Worsley Basin.

He also contrived and laid down within the mines
a system of underground railways, all leading from
the face of the coal, where the miners worked, to

the wells which he had made at different points
in the tunnels, through which the coals were shot
into the boats waiting below to receive them. At
Manchester, where they were unloaded for sale,
the contrivances which he employed were equally
ingenious. It was at first intended that the canal
should terminate at the foot of Castle Hill, up
which the coals were dragged by their purchasers
from the boats in wheelbarrows or carts. But the
toil of dragging the loads up the hill was found
very great; and, to remedy the inconvenience,
Brindley contrived to extend the canal for some
way into the hill, opening a shaft from the surface
of the ground down to the level of the water. The
barges having made their way to the foot of this
shaft, the boxes of coal were hoisted to the surface
by a crane, worked by a box water-wheel of 30 feet
diameter and 4 feet 4 inches wide, driven by the
waterfall of the river Medlock. In this contrivance Brindley was only adopting a modification
of the losing and gaining bucket, moved on a
vertical pillar, which he had before successfully
employed in drawing water out of coal-mines. By
these means the coals were rapidly raised to the
higher ground, where they were sold and distributed, greatly to the convenience of those who
came to purchase them.

Brindley's practical ability was equally displayed
in planning and building a viaduct or in fitting up
a crane—in carrying out an embankment or in contriving a coal-barge. The range and fertility of his
constructive genius were extraordinary. For the
Duke, he invented water-weights at Rough Close,
riddles to wash coal for the forges, raising dams,
and numerous other contrivances of well-adapted

mechanism. At Worsley he erected a steam-engine
for draining those parts of the mine which were
beneath the level of the canal, and consequently
could not be drained into it ; and he is said to have
erected, at a cost of only 150l., an engine which
until that time no one had known how to construct
for less than 500l. At the mouth of one of the
mines he erected a water-bellows for the purpose
of forcing fresh air into the interior, and thus ven-
tilating the workings.* At the entrance of the
underground canal he designed and built a mill of
a new construction, driven by an over-shot wheel
twenty-four feet in diameter, which worked three
pairs of stones for grinding corn, besides a dressing
or boulting mill, and a machine for sifting sand and
mixing mortar.

Brindley's quickness of observation and readiness in turning circumstances to advantage were
equally displayed in the mode by which he contrived to obtain an ample supply of lime for building purposes during the progress of the works.
We give the account as related by Arthur Young :—
"In carrying on the navigation," he observes, "a
vast quantity of masonry was necessary for building

---

* A writer in the 'St. James's Chronicle,' under date the 30th
of September, 1763, gives the following account of this apparatus,
long since removed :—"At the mouth of the cavern is erected a
water-bellows, being the body of a tree, forming a hollow cylinder,
standing upright. Upon this a wooden basin is fixed, in the form
of a funnel, which receives a current of water from the higher
ground. This water falls into the cylinder, and issues out at the
bottom of it, but at the same time carries a quantity of air with it,
which is received into the pipes and forced to the innermost
recesses of the coalpits, where it issues out as if from a pair of
bellows, and rarefies the body of thick air, which would otherwise
prevent the workmen from subsisting on the spot where the coals
are dug."

aqueducts, bridges, warehouses, wharves, &c., and
the want of lime was felt severely. The search
that was made for matters that would burn into
lime was for a long time fruitless. At last Mr.
Brindley met with a substance of a chalky kind,
which, like the rest, he tried; but found (though it
was of a limestone nature—lime-marl, which was
found along the sides of the canal, about a foot
below the surface) that, for want of adhesion in
the parts, it would not make lime. This most inventive genius happily fell upon an expedient to
remedy this misfortune. He thought of tempering
this earth in the nature of brick-earth, casting it in
moulds like bricks, and then burning it; and the
success was answerable to his wishes. In that
state it burnt readily into excellent lime; and this
acquisition was one of the most important that
could have been made. I have heard it asserted
more than once that this stroke was better than
twenty thousand pounds in the Duke's pocket;
but, like most common assertions of the same kind,
it is probably an exaggeration. However, whether
the discovery was worth five, ten, or twenty
thousand, it certainly was of noble use, and forwarded all the works in an extraordinary manner." *

It has been stated that Brindley's nervous excitement was so great on the occasion of the letting
of the water into the canal, that he took to his bed
at the Wheatsheaf, in Stretford, and lay there until
all cause for apprehension was over. The tension

* 'Six Months' Tour,' vol. iii., pp. 270–1. Mr. Hughes, C.E.,
says of this discovery : " The lime thus made would appear to be
the first cement of which we have any knowledge in this country ;
since the calcareous marl here spoken of would probably produce,
when burnt, a lime of strong hydraulic properties."

on his brain must have been great, with so tremendous a load of work and anxiety upon him; but that he "ran away,"* as some of his detractors have alleged, is at variance with the whole character and history of the man.

The Duke's canal, when finished, was for a long time regarded as the wonder of the neighbourhood. Strangers flocked from a distance to see Brindley's "castle in the air;" and contemporary writers spoke in glowing terms of the surprise with which they saw several barges of great burthen drawn by a single mule or horse along "a river hung in the air," over another river flowing underneath, by the

---

* This story was first set on foot, we believe, by the Earl of Bridgewater, in his singularly incoherent publication entitled, 'A Letter to the Parisians and the French Nation upon Inland Navigation, containing a defence of the public character of His Grace Francis Egerton, late Duke of Bridgewater. By the Hon. Francis Henry Egerton.' The first part of this curious book (published at Paris) was dated "Hôtel Egerton, Paris, 21st Dec., 1818;" the second part was published two years later; and a third part, consisting entirely of a note about Hebrew interpretations, was published subsequently. He had in the mean time become Earl of Bridgewater, in October, 1823, having formerly been prebendary of Durham and rector of Whitchurch in Shropshire. The late Earl of Ellesmere, in his 'Essays on History, Biography,' &c., says of this nobleman that "he died at Paris in the odour of eccentricity." But this is a mild description of his lordship, who had at least a dozen distinct crazes—about canals, the Jews, punctuation, the wonderful merits of the Egertons, the proper translation of Hebrew, the ancient languages generally, but more especially about prophecy and poodle-dogs. When he drove along the Boulevards in Paris, nothing could be seen of his lordship for poodle-dogs looking out of the carriage windows. The poodles sat at table with him at dinner, each being waited on by a special valet. The most creditable thing the Earl did was to leave the sum of 12,000l. to the British Museum, and 8000l. to meritorious literary men for writing the well-known 'Bridgewater Treatises.' He died in February, 1829.

side of which some ten or twelve men might be seen slowly hauling a single barge against the stream. A lady who writes a description of the work in 1765, speaks of it as "perhaps the greatest artificial curiosity in the world;" and she states that "crowds of people, including those of the first fashion, resort to it daily."

The chief importance of the work, however, consisted in its valuable uses. Manchester was now regularly and cheaply supplied with coals. The average price was at once reduced by one-half— from 7*d.* the cwt. to 3½*d.* (six score being given to the cwt.)—and the supply was regular instead of intermitting, as it had formerly been. But the full advantages of this improved supply of coals were not experienced until many years after the opening of the canal, when the invention of the steam-engine, and its extensive employment as a motive power in all manufacturing operations, rendered a cheap and abundant supply of fuel of vital importance to the growth and prosperity of Manchester and its neighbourhood.

CHAPTER IV.

EXTENSION OF THE DUKE'S CANAL TO THE MERSEY.

THE CANAL had scarcely been opened to Manchester when we find Brindley occupied, at the instance of the Duke, in surveying the country between Stretford and the river Mersey, with the view of carrying out a canal in that direction for the accommodation of the growing trade between Liverpool and Manchester. The first boat-load of coals sailed over the Barton viaduct to Manchester on the 17th of July, 1761 ; on the 7th of September following we find Brindley at Liverpool,* " rocconitoring ; " and, by the end of the month, he was busily engaged levelling for a proposed canal to join the Mersey at

---

* It would almost seem as if the extension of the canal to the Mersey had formed part of the Duke's original plan ; for Brindley was engaged in making a survey from Longford to Dunham in the autumn of the preceding year, as appears from the following account, preserved at the Bridgewater Canal Office at Manchester, of his expenses in making the survey :—

" Expenses in Surveying from Longford Bridge to Dunham.

Octr 21st 1760.

Spent at Stretford	... 0	6
At Altringham all Night	... 6	0
Gave the Men to Drink that assisted	... 1	0
22nd		
More at Altringham	... 2	6
	10	0

P^d Mr. Brinley this."

Hempstones, about eight miles below Warrington Bridge, from whence there was a natural tideway to Liverpool, about fifteen miles distant.

The project in question was a very important one on public grounds. We have seen how the community of Manchester had been hampered by defective road and water communications, which seriously affected its supplies of food and fuel, and, at the same time, by retarding its trade, hindered to a considerable extent the regular employment of its population. The Duke of Bridgewater, by constructing his canal, had opened up an abundant supply of coal, but the transport of the raw materials of manufacture was still as much impeded as before. Liverpool was the natural port of Manchester, from which it drew its supplies of cotton, wool, silk, and other produce, and to which it returned them for export when worked up into manufactured articles.

There were two existing modes by which the communication was kept up between the two places : one was by the ordinary roads, and the other by the rivers Mersey and Irwell. From a statement published in December, 1761, it appears that the weight of goods then carried by land from Manchester to Liverpool was "upwards of forty tons per week," or about two thousand tons a year. This quantity, insignificant though it must appear when compared with the enormous traffic now passing between the two towns, was then thought very large, as no doubt it was when the limited trade of the country is taken into account. But the cost of transport was the important feature ; it was not less than two pounds sterling per ton— this heavy charge being almost entirely attributable to the execrable state of the roads. It was scarcely possible to drive waggons along the ruts and through the sloughs which lay between the two places at certain seasons of the year, and even pack-horses had considerable difficulty in making the journey.

The other route between the towns was by the navigation of the rivers Mersey and Irwell. The raw materials used in manufacture were principally transported from Liverpool to Manchester by this route, at a cost of about twelve shillings per ton ; the carriage of timber and such like articles costing not less than twenty per cent. on their value at Liverpool. But the navigation was also very tedious and difficult. The boats could only pass up to the first lock at the Liverpool end with the assistance of a spring tide; and further up the river there were numerous fords and shallows which the boats could only pass in great freshes, or in dry seasons, by drawing extraordinary quantities of water from the locks above. Then, in winter, the navigation was apt to be impeded by floods, and occasionally it was stopped altogether. In short, the growing wants of the population demanded an improved means of transit between the two towns, which the Duke of Bridgewater now determined to supply.

The growth of Liverpool as a seaport has been comparatively recent. At a time when Bristol and Hull possessed thriving harbours, resorted to by foreign ships, Liverpool was little better than a fishing village, its only distinction being that it was a convenient place for setting sail to Ireland. In the war between France and England which broke out in 1347, when Edward the Third summoned the various ports in the kingdom to make contributions towards the naval power according to their means, London was required to provide 25 ships and 662 men; Bristol 22 ships and 608 men; Hull, 16 ships and 466 men; whilst Liverpool was only asked to find 1 bark and 6 men! In Queen Elizabeth's time, the burgesses presented a petition to Her Majesty, praying her to remit a subsidy which had been imposed upon it and other seaport towns, in which they styled their native place "Her Majesty's poor decayed town of Liverpool." Chester was then of considerably greater importance as a port. In 1634–5, when Charles I. made his unconstitutional levy of ship-money throughout England, Liverpool was let off with a contribution of 15*l.*, whilst Chester paid 100*l.*, and Bristol not less than 1000*l.*

The channel of the Dee, however, becoming silted up, the trade of Chester decayed, and that of Liverpool rose upon its ruins. In 1699 the excavation of the Old Dock was begun; but it was used only as a tidal harbour (being merely an enclosed space with a small pier) until the year 1709, when an Act was obtained enabling its conversion into a wet dock; since which time a series of docks have been constructed, extending for about five miles along the north shore of the Mersey, which are among the greatest works of modern times, and afford an almost unequalled amount of shipping accommodation.

From that time forward the progress of the port of Liverpool has kept steady pace with the trade and wealth of the country behind it, and especially with the manufacturing activity and energy of the town of Manchester. Its situation at the mouth of a deep and navigable river, its convenient proximity to districts abounding in coal and iron and inhabited by an industrious and hardy population, were unquestionably great advantages. But these of themselves would have been insufficient to account for the extraordinary progress made by Liverpool during the last century, without the opening up of the great system of canals, which brought not only the towns of Yorkshire, Cheshire, and Lancashire into immediate connection with that seaport, but also the manufacturing districts of Staffordshire, Warwickshire, and the other central counties of England situated at the confluence of the various navigations.* Liverpool thus became the great focus of import and export for the northern and western districts. The raw materials of commerce were poured into it from Ireland, America, and the Indies. From thence they were distributed along the canals amongst the various seats of manufacturing industry, and a large proportion was readily returned by the same route to the same port, in a manufactured state, for shipment to all parts of the world.

At the time of which we speak, however, it will be observed that the communication between Liverpool and Manchester was very imperfect. It was not only difficult to convey goods between the two places, but it was also difficult to convey persons. In fine weather, those who required to travel the thirty miles which separated them, could ride or walk, resting at Warrington for the night.

* Progress of Liverpool.

Years.	Vessels entered.	Tonnage.	Duties paid.
1701	102	8,619	...
1760	1,245	...	£2,330
1800	4,746	450,060	23,379
1858	21,352	4,441,943	347,889

But in winter the roads, like most of the other
country roads at the time, were simply impassable.
Although an Act had been passed as early as the
year 1726 for repairing and enlarging the road
from Liverpool to Prescot, coaches could not come
nearer to the town than Warrington in 1750, the

Liverpool in 1650.

road being impracticable for such vehicles even in
summer.*

A stage-coach was not started between Liverpool and Manchester until the year 1767, performing
the journey only three times a week.   It required

* Mr. Baines says : "Carriages were then very rare, and it is
mentioned as a singular fact that at the period in question (1750)
there was but one *gentleman's* carriage in the town of Liverpool,
and that carriage was kept by a *lady* of the name of Clayton."—
'History of Lancashire,' vol. iv., p. 90.

six and sometimes eight horses to draw the lumbering vehicle and its load along the ruts and through the sloughs, — the whole day being occupied in making the journey. The coach was accustomed to start early in the morning from Liverpool; it breakfasted at Prescot, dined at Warrington, and arrived at Manchester usually in time for supper. On one occasion, at Warrington, the coachman intimated his wish to proceed, when the company requested him to take another pint, as they had not finished their wine, asking him at the same time if he was in a hurry? "Oh," replied the driver, "I'm not partic'lar to an hour or so!" As late as 1775, no mail-coach ran between Liverpool and any other town, the bags being conveyed to and from it on horseback; and one letter-carrier was found sufficient for the wants of the place. A heavy stage then ran, or rather crawled, between Liverpool and London, making only four journeys a week in the winter time. It started from the Golden Talbot, in Water-street, and was three days on the road. It went by Middlewich, where one of its proprietors kept the White Bear inn; and during the Knutsford race-week the coach was sent all the way round by that place, in order to bring customers to the Bear.

We have said that Brindley was engaged upon the preliminary survey of a canal to connect Manchester with the Mersey, immediately after the original Worsley line had been opened, and before its paying qualities had been ascertained. But the Duke, having once made up his mind as to the expediency of carrying out this larger project, never halted nor looked back, but made arrangements for prosecuting a bill for the purpose of enabling

the canal to be made in the very next session of
Parliament.

We find that Brindley's first visit to Liverpool
and the intervening district on the business of the
survey was made early in September, 1761. During
the remainder of the month he was principally
occupied in Staffordshire, looking after the working of his fire-engine at Fenton Vivian, carrying
out improvements in the silk-manufactory at Congleton, and inspecting various mills at Newcastle-
under-Lyne and the neighbourhood. His only idle
day during that month seems to have been the 22nd,
which was a holiday, for he makes the entry in his
book of "crounation of Georg and Sharlot," the new
King and Queen of England. By the 25th we find
him again with the Duke at Worsley, and on the
30th he makes the entry, "set out at Dunham to
Level for Liverpool." The work then went on
continuously until the survey was completed; and
on the 19th of November he set out for London,
with 7*l.* 18*s.* in his pocket.

In the course of his numerous journeys, we
find Brindley carefully noting down the various
items of his expenses, which were curiously small.
Although he was four or five days on the road to
London, and stayed eight days there, his total expenses, both going and returning, amounted to
only 4*l.* 8*s.* : it is most probable, however, that he
lived at the Duke's house whilst in town. On the
1st of December we find him, on his return journey
to Worsley, resting the first night at a place called
Brickhill; the next at Coventry, where he makes
the entry, "Moy mar had a bad fall in the frasst;"
the third at Sandon; the fourth at Congleton; and
the fifth at Worsley. He had still some inquiries

to make as to the depth of water and the conditions of the tide at Hempstones; and for three days he seems to have been occupied in traffic-taking, with a view to the evidence to be given before Parliament; for on the 10th of December we find him at Stretford, "to count the caridgos," and on the 12th he is at Manchester for the same purpose, "counting the loded caridgos and horses."

The following bill refers to some of the work done by him at this time, and is a curious specimen of an engineer's travelling charges in those days— the engineer himself being at the same time paid at the rate of 3s. 6d. a day :—

Expenses for His Grace the Duk of Bridgwater to pay for traveling Chareges by James Brindley.

18 Novem—1761.

18 No masuring a Cros from Dunham to Warburton Mercey and Thalwall, 3s - 11d Dunham for 2 diners 1s - 3d for the man 1s - 0d at Thalwall 1s - 2d all Night Warington 0 7 4

19 Novem Sat out from Chester for London & at Worsley Septm 5 Retorned back going to London and at London and hors back to Worsley Charged Hors & my salf 4 8 0

9 december Coming back from Ham Stone Charges at Wilderspool all Night 0 8 0

at Warington to meet Mr Ashley   dining   ... 0 4 2

10 to ataind the Turn pike Rode 2s - 6d & againe on te 12 De Rode 3s - 6d 0 6 0

21 Decm to inspect te flux and Reflux at Ham Stone 2 dayes Charges 0 6 6
_____

26 Decr 1761. Recd the Contents of the above Bill by the Hands of John Gilbert. James Brindley   ... £6 00 0

In the early part of the month of January, 1762, we find Brindley busy measuring soughs, gauging

the tides at Hempstones, and examining and altering the Duke's paper-mills and iron slitting-mills at
Worsley; and on the 7th we find this entry: "to
masuor the Duks pools I and Smeaton." On the
following day he makes "an ochilor survey from
Saldnoor [Sale Moor] to Stockport," with a view
to a branch canal being carried in that direction.
On the 14th, he sets out from Congleton, by way of
Ashbourne, Northampton, and Dunstable, arriving
in London on the fifth day.

Immediately on his arrival in town we find him
proceeding to rig himself out in a new suit of
clothes. His means were small, his habits thrifty,
and his wardrobe scanty; but as he was about to
appear in an important character, as the principal
engineering witness before a Parliamentary Committee in support of the Duke's bill, he felt it
necessary to incur an extra expenditure on dress
for the occasion. Accordingly, on the morning of
the 18th we find him expending a guinea—an entire
week's pay — in the purchase of a pair of new
breeches: two guineas on a coat and waistcoat of
broadcloth, and six shillings for a pair of new
shoes. On the following page is a fac-simile of the
entry in his pocketbook.

It will be observed that an expenditure is here
entered of nine shillings for going to "the play."
It would appear that his friend Gilbert, who was
in London with him on the canal business, prevailed on Brindley to go with him to the theatre to
see Garrick in the play of 'Richard III.,' and he
went. He had never been to an entertainment of
the kind before; but the excitement which it caused
him was so great, and it so completely disturbed
his ideas, that he was unfitted for business for

several days after. He then declared that no consideration should tempt him to go a second time,
and he held to his resolution. This was his first
and only visit to the play. The following week he

Fac-simile of Brindley's Hand-writing.

enters in his memorandum-book concerning himself
"ill in bed," and the first Sunday after his recovery
we find him attending service at "Sant Mary's
Church." The service did not make him ill, as the

play had done, and on the following day he attended the House of Commons on the subject of the Duke's bill.

The proposed canal from Manchester to the Mersey at Hempstones stirred up an opposition which none of the Duke's previous bills had encountered. Its chief opponents were the proprietors of the Mersey and Irwell Navigation, who saw their monopoly assailed by the measure; and, unable though they had been satisfactorily to conduct the then traffic between Liverpool and Manchester, they were unwilling to allow of any additional water service being provided between the two towns. Having already had sufficient evidence of the Duke's energy and enterprise, from what he had been able to effect in so short a time in forming the canal between Worsley and Manchester, the Navigation Company were not without reason alarmed at his present project.

At first they tried to buy him off by concessions. They offered to reduce the rate of 3s. 4d. per ton of coals, timber, &c., conveyed upon the Irwell between Barton and Manchester, to 6d. if he would join their Navigation at Barton and abandon the part of his canal between that point and Manchester; but he would not now be diverted from his plan, which he resolved to carry into execution if possible. Again they tried to conciliate his Grace by offering him certain exclusive advantages in the use of their Navigation. But it was again too late; and the Duke having a clear idea of the importance of his project, and being assured by his engineer of its practicability and the great commercial value of the undertaking, determined to proceed with the measure. It offered

to the public the advantages of a shorter line of
navigation, not liable to be interrupted by floods
on the one hand or droughts on the other, and,
at the same time, a much lower rate of freight,
the maximum charge proposed in the bill being
6s. a ton against 12s., the rate charged by the
Mersey and Irwell Navigation between Liverpool
and Manchester.

The opposition to the bill was led by Lord
Strange, son of the Earl of Derby, one of the
members for the county of Lancaster, who took
the part of the "Old Navigators," as they were
called, in resisting the bill. The question seems
also to have been treated as a political one; and,
the Duke and his friends being Whigs, Lord Strange
mustered the Tory party strongly against him.
Hence we find this entry occurring in Brindley's
note-book, under date the 16th of February: "The
Toores [Tories] mad had [made head] agane ye
Duk."

The principal objections offered to the proposed
canal were, that the landowners would suffer by
it from having their lands cut through and covered
with water, by which a great number of acres
would be for ever lost to the public; that there
was no necessity whatever for the canal, the Mersey
and Irwell Navigation being sufficient to carry
more goods than the trade then existing required;
that the new navigation would run almost parallel
with the old one, and offered no advantage to the
public which the existing river navigation did not
supply; that the canal would drain away the waters
which supplied the rivers, and be very prejudicial
to them, if not totally destructive, in dry seasons;
that the proprietors of the old navigation had

invested their money on the faith of protection by Parliament, and to permit the new canal to be established would be a gross interference with their vested rights; and so on.

To these objections there were very sufficient answers. The bill provided for full compensation being made to the owners of lands through which the canal passed, and, in addition, it was provided that all sorts of manure should be carried for them without charge. It was also shown that the Duke's canal could not abstract water from either the Mersey or the Irwell, as the level of both rivers was considerably below that of the intended canal, which would be supplied almost entirely from the drainage of his own coal-mines at Worsley; and with respect to the plea of vested rights set up, it was shown that Parliament, in granting certain powers to the old navigators, had regard mainly to the convenience and advantage of the public; and they were not precluded from empowering a new navigation to be formed if it could be proved to present a more convenient and advantageous mode of conveyance.

On these grounds the Duke was strongly supported by the inhabitants of the localities proposed to be served by the intended canal. The "Junto of Old Navigators of the Mersey and Irwell Company" had for many years carried things with a very high hand, extorted the highest rates, and, in cases of loss by delay or damage to goods in transit, refused all redress. A feeling very hostile to them and their monopoly had accordingly grown up, which now exhibited itself in a powerful array of petitions to Parliament in favour of the Duke's bill.

On the 17th of February, 1762, the bill came before the Committee of the House of Commons, and Brindley was examined in its support. We regret that no copy of his evidence now exists * from which we might have formed an opinion of the engineer's abilities as a witness. Some curious anecdotes have, however, been preserved of his demeanour and evidence on canal bills before Parliament. When asked, on one occasion, to produce a drawing of an intended bridge, he replied that he had no plan of it on paper, but he would illustrate it by a model. He went out and bought a *large cheese*, which he brought into the room and cut into two equal parts, saying, "Here is my model." The two halves of the cheese represented the semicircular arches of his bridge; and by laying over them some long rectangular object, he could thus readily communicate to the Committee the position of the river flowing underneath and the canal passing over it.†

On another occasion, when giving his evidence, he spoke so frequently about "puddling," describing its uses and advantages, that some of the members expressed a desire to know what this extraordinary mixture was, that could be applied to such important purposes. Preferring a practical illustration to a verbal description,

* Search has been made at the Bridgewater Estate Offices at Manchester, and in the archives of the Houses of Parliament, but no copy can be found. It is probable that the Parliamentary papers connected with this application to Parliament were destroyed by the fire which consumed so many similar documents about twenty-five years ago.

† Stated by Mr. Hughes, in his 'Memoir of Brindley,' as having been communicated to him by James Loch, Esq., M.P., formerly agent for the Duke's Trustees.

Brindley caused a mass of clay to be brought into the committee-room, and, moulding it in its raw untempered state into the form of a trough, he poured into it some water, which speedily ran through and disappeared. He then worked the clay up with water to imitate the process of puddling, and again forming it into a trough, filled it with water, which was now held in without a particle of leakage. "Thus it is," said Brindley, "that I form a water-tight trunk to carry water over rivers and valleys, wherever they cross the path of the canal." *

Again, when Brindley was giving evidence before a Committee of the House of Peers as to the lockage of his proposed canal, one of their Lordships asked him, "But what is a lock?" on which the engineer took a piece of chalk from his pocket and proceeded to explain it by means of a diagram which he drew upon the floor, and made the matter clear at once.† He used to be so ready with his chalk for purposes of illustration, that it became a common saying in Lancashire, that "Brindley and chalk would go through the world." He was never so eloquent as when with his chalk in hand,—it stood him in lieu of tongue.

---

* 'Memoir of Brindley,' by S. Hughes, C.E., in 'Weale's Papers on Civil Engineering.'

† As the reader may possibly desire information on the same point, we may here briefly explain the nature of a Canal Lock. It is employed as a means of carrying navigations through an uneven country, and raising the boats from one water level to another, or *vice versâ*. The lock is a chamber formed of masonry, occupying the bed of the canal where the difference of level is to be overcome. It is provided with two pairs of gates, one at each end ; and the chamber is so contrived that the level of the water which

On the day following Brindley's examination
before the Committee on the Duke's bill, that is,
on the 18th of February, we find him entering in
his note-book that the Duke sent out " 200 leators "

it contains may be made to coincide with either the higher level
above, or the lower level below it. The following diagrams will
explain the form and construction of the lock. A represents what
is called the upper pond, B the lower, C is the left wall, and DD
side culverts. When the gates at the lower end of the chamber
(E) are opened, and those at the upper end (F) are closed, the

Longitudinal Section and Plan of Lock.

water in the chamber will stand at the lower level of the canal ;
but when the lower gates are closed, and the upper gates are
opened, the water will naturally coincide with that in the upper
part of the canal. In the first case, a boat may be floated into the
lock from the lower part, and then, if the lower gates be closed
and water is admitted from the upper level, the canal-boat is
raised, by the depth of water thus added to the lock, to the upper
level, and on the complete opening of the gates it is thus floated
onward. By reversing the process, it will readily be understood
how the boat may, in like manner, be lowered from the higher to

to members — possible friends of the measure;
containing his statement of reasons in favour of
the bill. On the 20th Mr. Tomkinson, the Duke's
solicitor, was under examination for four hours
and a half. Sunday intervened, on which day
Brindley records that he was "at Lord Harring-
ton's." On the following day, the 22nd, the evidence for the bill was finished, and the Duke
followed this up by sending out 250 more letters
to members, with an abstract of the evidence
given in favour of the measure. On the 26th
there was a debate of eight hours on the bill,
followed by a division, in Committee of the whole
House, thus recorded by Brindley :—

> "ad a grate Division of 127 fort Duk
>   98 nos
>   —
>   for t⁰ Duk 29 Me Jorete"

But the bill had still other discussions and
divisions to encounter before it was safe. The
Duke and his agents worked with great assiduity.
On the 3rd of March he caused 250 more letters
to be distributed amongst the members ; and on
the day after we find the House wholly occupied
with the bill. We quote again from Brindley's

the lower level. The greater the lift or the lowering, the more
water is consumed in the process of exchange from one level to
another ; and where the traffic of the canal is great, a large supply
of water is required to carry it on, which is usually provided by
capacious reservoirs situated above the summit level. Various
expedients are adopted for economising water : thus, when the
width of the canal will admit of it, the lock is made in two com-
partments, communicating with each other by a valve, which can
be opened and shut at pleasure ; and by this means one-half of
the water which it would otherwise be necessary to discharge to
the lower level may be transferred to the other compartment.

record: "4 [March] ade bate at the Hous with grate vigor 3 divisons the Duke carred by Numbers evory time a 4 division moved but Noos yelded." On the next day we read "wont thro the closos;" from which we learn that the clauses were settled and passed. Mr. Gilbert and Mr. Tomkinson then set out for Lancashire: the bill was safe. It passed the third reading, Brindley making mention that "Lord Strange" was "sick with geef [grief] on that affair Mr. Wellbron want Rong god,"—which latter expression we do not clearly understand, unless it was that Mr. Wilbraham wanted to wrong God. The bill was carried to the Lords, Brindley on the 10th March making the entry, "Touk the Lords oath." But the bill passed the Upper House "without opposishin," and received the Royal Assent on the 24th of the same month.

On the day following the passage of the bill through the House of Lords (of which Brindley makes the triumphant entry, "Lord Strange defetted"), he set out for Lancashire, after nine weary weeks' stay in London. To hang about the lobbies of the House and haunt the office of the Parliamentary agent, must have been excessively irksome to a man like Brindley, accustomed to incessant occupation and to see works growing under his hands. During this time we find him frequently at the office of the Duke's solicitor in "Mary Axs;" sometimes with Mr. Tomkinson, who paid him his guinea a-week during the latter part of his stay; and on several occasions he is engaged with gentlemen from the country, advising them about "saltworks at Droitwitch" and mill-arrangements in Cheshire.

I.   R

Many things had fallen behind during his absence and required his attention, so he at once set out home; but the first day, on reaching Dunstable, he was alarmed to find that his mare, so long unaccustomed to the road, had "allmost lost ye use of her Limes" [limbs]. He therefore went on slowly, as the mare was a great favourite with him—his affection for the animal having on one occasion given rise to a serious quarrel between him and Mr. Gilbert—and he did not reach Congleton until the sixth day after his setting out from London. He rested at Congleton for two days, during which he "settled the geering of the silk-mill," and then proceeded straight on to Worsley to set about the working survey of the new canal.

The course of this important canal, which unites the mills of Manchester with the shipping of Liverpool, is about twenty-four miles in length.* From Longford Bridge, near Manchester, its course lies in a south-westerly direction for some distance, crossing the river Mersey at a point about five miles above its junction with the Irwell. At Altrincham it proceeds in a westerly direction, crossing the river Bollin about three miles further on, near Dunham. After crossing the Bollin, it describes a small semicircle, proceeding onward

---

* The following statement of the lengths of the different portions of the Duke's canal, including those originally executed, is from the map published by Brindley in 1769 :—

	Miles.	furl.	chains.	
From Worsley to Longford Bridge	6	0	0	Level.
,, Longford Bridge to Manchester	4	2	0	,,
,, Longford Bridge to Preston Brook	19	0	0	,,
,, Preston Brook to upper part of Runcorn	4	4	0	,,
,, Upper part of Runcorn to the Mersey	0	5	7	79 feet fall.

in the valley of the Mersey, and nearly in the direction of the river as far as the crossing of the high road from Chester to Warrington. It then bends to the south to preserve the high level, passing in a southerly direction as far as Preston, in Cheshire, from whence it again turns round to the north to join the river Mersey. [For Map of the Canal, see pp. 204-5.]

The canal lies entirely in the lower part of the new red sandstone, the principal earthworks consisting of the clays, marls, bog-earths, and occasionally the sandstones of this formation. The heaviest bog crossed in the line of the works was Sale Moor, west of the Mersey, where the bottom was of quicksand; and the construction of the canal at that part was probably an undertaking of as formidable a character as the laying of the railroad over Chat Moss proved some sixty years later. But Brindley, like Stephenson, looked upon a difficulty as a thing to be overcome; and when an obstruction presented itself, he at once set his wits to work and studied how it was best to be grappled with and surmounted. A large number of brooks had to be crossed, and also two important rivers, which involved the construction of numerous aqueducts, bridges, and culverts, to provide for the surface water supply of the district. It will, therefore, be obvious that the undertaking was of a much more important nature—more difficult for the engineer to execute, and more costly to the noble proprietor who found the means for carrying it to a completion—than the comparatively limited and inexpensive work between Worsley and Manchester, which we have above described.

The capital idea which Brindley early formed

and determined to carry out, was to construct a
level of dead water all the way from Manchester
to a point as near to the junction of the canal with
the Mersey as might be found practicable. Such
a canal, he clearly saw, would not be so expensive
to work as one furnished with locks at intermediate
points. Brindley's practice of securing long levels
of water in canals was in many respects similar to
that of George Stephenson with reference to flat
gradients upon railways; and in all the canals that
he constructed, he planned and carried them out
as far as possible after this leading principle.
Hence the whole of the locks on the Duke's canal
were concentrated at its lower end near Runcorn,
where the navigation descended, as it were by a
flight of water steps, into the river Mersey. Lord
Ellesmere has observed that this uninterrupted
level of the Bridgewater Canal from Leigh and
Manchester to Runcorn, and the concentration of
its descent to the Mersey at the latter place, have
always been considered as among the most striking
evidences of the genius and skill of Brindley.

There was, as usual, considerable delay in
obtaining possession of the land on which to
commence the works. The tenants required a
certain notice, which must necessarily expire before
the Duke's engineer could take possession; and
numerous obstacles were thrown in his way, both
by tenants and landlords hostile to the undertaking.
In many cases the Duke had to pay dearly for the
land purchased under the compulsory powers of
his Act. Near Lymm, the canal passed through a
little bit of garden belonging to a poor man's cottage, the only produce growing upon the ground
being a pear-tree. For this the Duke had to pay

thirty guineas, and it was thought a very extravagant price at that time. Since the introduction of railways, the price would probably be considered ridiculously low. For the land on which the warehouses and docks were built at Manchester, the Duke had to pay in all the much more formidable sum of about forty thousand pounds.

The Old Quay Navigation, even at the last moment, thought to delay if not to defeat the Duke's operations, by lowering their rates nearly one-half. Only a few days after the Royal Assent had been given to the bill, they published an announcement, appropriately dated the 1st of April, setting forth the large sacrifices they were about to make, and intimating that " from their Reductions in Carriage a real and permanent Advantage will arise to the Public, and they will experience that Utility so cried up of late, but which has hitherto only existed in promises." The Duke heeded not the ineffective blow thus aimed at him : he was only more than ever resolved to go forward with his canal. He was even offered the Mersey Navigation itself at the price of thirteen thousand pounds ; but he would not have it now at any price.

The public spirit and enterprise displayed by many of the young noblemen of those days was truly admirable. Brindley had for several years been in close personal communication with Earl Gower as to the construction of the canal intended to unite the Mersey with the Trent and the Severn, and thus connect the ports of Liverpool, Hull, and Bristol, by a system of inland water-communication. With this object, as we have seen, he had often visited the Earl at his seat at Trentham, and discussed with him the plans by which this truly

magnificent enterprise was to be carried out ; and
he had frequently visited the Earl of Stamford at
his seat at Enville for the same purpose. But those
schemes were too extensive and costly to be carried
out by the private means of either of those noblemen, or even by both combined. They were,
therefore, under the necessity of stirring up the
latent enterprise of the landed proprietors in their
respective districts, and waiting until they had
received a sufficient amount of local support to
enable them to act with vigour in carrying their
great design into effect.

The Duke of Bridgewater's scheme of uniting
Manchester and Liverpool by an entirely new line
of water-communication, cut across bogs and out
of the solid earth in some places, and carried over
rivers and valleys at others by bridges and embankments, was scarcely less bold or costly. Though
it was spoken of as another of the Duke's " castles
in the air," and his resources were by no means
overflowing at the time he projected it, he nevertheless determined to go on alone with it, should no
one be willing to join him. The Duke thus proved
himself a real Dux or leader of industrial enterprise
in his district ; and by cutting his canal, and providing a new, short, and cheap water-way between
Liverpool and Manchester, which was afterwards
extended through the counties of Chester, Stafford,
and Warwick, he unquestionably paved the way
for the creation and development of the modern
manufacturing system existing in the north-western
counties of England.

We need scarcely say how admirably he was
supported throughout by the skill and indefatigable
energy of his engineer. Brindley's fertility in

resources was the theme of general admiration. Arthur Young, who visited the works during their progress, speaks with enthusiastic admiration of his "bold and decisive strokes of genius," his "penetration which sees into futurity, and prevents obstructions unthought of by the vulgar mind, merely by foreseeing them: a man," says he, "with such ideas, moves in a sphere that is to the rest of the world imaginary, or at best a *terra incognita.*"

It would be uninteresting to describe the works of the Bridgewater Canal in detail; for one part of a canal is usually so like another, that to do so were merely to involve a needless amount of repetition of a necessarily dry description. We shall accordingly content ourselves with referring to the original methods by which Brindley contrived to overcome the more important difficulties of the undertaking.

From Longford Bridge, where the new works commenced, the canal, which was originally about eight yards wide and four feet deep, was carried upon an embankment of about a mile in extent across the valley of the Mersey. One might naturally suppose that the conveyance of such a mass of earth must have exclusively employed all the horses and carts in the neighbourhood for years. But Brindley, with his usual fertility in expedients, contrived to make the construction of one part of the canal subservient to the completion of the remainder. He had the stuff required to make up the embankment brought in boats partly from Worsley and partly from other parts of the canal where the cutting was in excess ; and the boats, filled with this stuff, were conducted from

the canal along which they had come into watertight
caissons or cisterns placed at the point over which
the earth and clay had to be deposited.

The boats, being double, fixed within two feet
of each other, had a triangular trough supported
between them of sufficient capacity to contain about
seventeen tons of earth. The bottom of this trough
consisted of a line of trap-doors, which flew open
at once on a pin being drawn, and discharged their
whole burthen into the bed of the canal in an
instant. Thus the level of the embankment was
raised to the point necessary to enable the canal
to be carried forward to the next length. Arthur
Young was of opinion that the saving effected by

Brindley's Ballast Boats.

constructing the Stretford embankment in this
way, instead of by carting the stuff, was equivalent
to not less than five thousand per cent.! The
materials of the caissons employed in executing
this part of the work were afterwards used in
forming temporary locks across the valley of the
Bollin, whilst the embankment was being constructed at that point by a process almost the very
reverse, but of equal ingenuity.

In the same valley of the Mersey the canal
had to be carried over a large brook subject to
heavy floods, by means of a strong bridge of two
arches, adjoining which was a third, affording provision for a road. Further on, the canal was

carried over the Mersey itself upon a bridge with one arch of seventy feet span. Westward of this river lay a very difficult part of the work, occasioned by the carrying of the navigation over the Sale Moor Moss. Many thought this an altogether impracticable thing; as not only had the hollow trunk of earth in which the canal lay to be made water-tight, but to preserve the level of the water-way it must necessarily be raised considerably above the level of the Moor across which it was to be laid. Brindley overcame the difficulty in the following manner. He made a strong casing of timber-work outside the intended line of embankment on either side of the canal, by placing deal balks in an erect position, backing and supporting them on the outside with other balks laid in rows, and fast screwed together; and on the front side of this woodwork he had his earth-work brought forward, hard rammed, and puddled, to form the navigable canal; after which the casing was moved onward to the part of the work further in advance, and the bottom having previously been set with rubble and gravel, the embankment was thus carried forward by degrees, the canal was raised to the proper level, and the whole was substantially and satisfactorily finished.

A steam-engine of Brindley's contrivance was erected at Dunham Town Bridge to pump the water from the foundations there. The engine was called a Sawney, for what reason is not stated, and, for long after, the bridge was called Sawney's Bridge. The foundations of the under-bridge, near the same place, were popularly supposed to be set on quicksand; and old Lord Warrington, when he had occasion to pass under

it, would pretend cautiously to look about him,
as if to examine whether the piers were all right,
and then run through as fast as he could. A tall
poplar-tree stood at Dunham Banks, on which a
board was nailed showing the height of the canal
level; the people long after called the place "The
Duke's Folly," the name given to it while his
scheme was still believed to be impracticable.
But the skill of the engineer baffled these and
other prophets of evil; and the success of his
expedients, in nearly every case of difficulty that
occurred, must certainly be regarded as remarkable, considering the novel and unprecedented
character of the undertaking.

Brindley invariably contrived to economise
labour as much as possible, and many of his expedients with this object were very ingenious.
So far as he could, he endeavoured to make use
of the canal itself for the purpose of forwarding
the work. He had a floating blacksmith's forge
and shop, provided with all requisite appliances,
fitted up in one barge; a complete carpenter's
shop in another; and a mason's shop in a third;
all of which were floated on as the canal advanced, and were thus always at hand to supply
the requisite facilities for prosecuting the operations with economy and despatch. Where there
was a break in the line of work, occasioned, for
instance, by the erection of some bridge not yet
finished, the engineer had similar barges constructed and carried by land to other lengths of
the canal which were in progress, where they
were floated and advanced in like manner for the
use of the workmen. When the bridge across
the Mersey, which was pushed on as rapidly as

possible with the object of economising labour and cost of materials, was completed, the stone, lime, and timber were brought along the canal from the Duke's property at Worsley, as well as supplies of clay for the purpose of puddling the bottom of the waterway; and thus the work rapidly advanced at all points.

As one of the great objections made to the construction of the canal had been the danger threatened to the surrounding districts by the bursting of the embankments, Brindley made it his object to provide against the occurrence of such an accident by an ingenious expedient. He had stops or flood-gates contrived and laid in various parts of the bed of the canal, across its bottom, so that, in the event of a breach occurring in the bank and a rush of water taking place, the current which must necessarily set in to that point should have the effect of immediately raising the valvular floodgates, and so shutting off the stream and preventing the escape of more water than was contained in the division between the two nearest gates on either side of the breach. At the same time, these flood-gates might be used for cutting off the waters of the canal at different points, for the purpose of making any necessary repairs in particular lengths; the contrivance of waste tubes and plugs being so arranged that the bed of any part of the canal, more especially where it passed over the bridges, might be laid bare in a few hours, and the repairs executed at once.

In devising these ingenious expedients, it ought to be remembered that Brindley had no previous experience to fall back upon, and possessed no knowledge of the means which foreign engineers might have adopted to meet similar emergencies. All had been the result of his own original thinking and contrivance ; and, indeed, many of these devices were altogether new and original, and had never before been tried by any engineer.

It is curious to trace the progress of the works by Brindley's own memoranda, which, though brief, clearly exhibit his marvellous industry and close application to every detail of the business. He settled with the farmers for their tenant-right, sold and accounted for the wood cut down and the gravel dug out along the line of the canal, paid the workmen employed,* laid out the work, measured off the quantities done from time to time, planned and erected the bridges, designed the canal boats required for conveying the earth to form the embankments, and united in himself the varied functions of land-surveyor, carpenter, mason, brickmaker, boat-builder, paymaster, and engineer. We find him even condescending to count bricks and sell grass. Nothing was too

---

* The following bill is preserved amongst the Bridgewater Canal papers. Simcox was a skilled mechanic, and acted as foreman of the carpenters :—

"His Grace the Duke of Bridgewater to Sam¹ Simcox.   Dr

		£.	s.	d.
23 Mar^h 1760	To 12 days work at 21^d per	1	1	0
23 Aug^t	To 6 days more d° at d°	0	10	6
6 Sep^r	To 8 days more d° at d°	0	14	0
		2	5	6

1 Novr 1760. Rec^d the Contents above by the Hands of John Gilbert for the Use of Sam¹ Simcox.   P^d

JAMES BRINDLEY."

The wages of what was called a "right-hand man" at that time were from 14d. to 16d. a day, and of a "left-hand man" from 1s. to 14d.

small for him to attend to, nor too bold for him to undertake, when necessity required. At the same time we find him contriving a water-plane for the Duke's collieries at Worsley, and occasionally visiting Newchapel, Leek, and Congleton, in Cheshire, for the purpose of attending to the business on which he still continued to be employed at those places.

The heavy works at the crossing of the Mersey occupied him almost exclusively towards the end of the year 1763. He was there making dams and pushing on the building of the bridge. Occasionally he enters the words, "short of men at Cornbrook." Indeed, he seems at that time to have lived upon the works, for we find the almost daily entry of "dined at the Bull, 8*d*." On the 10th of November he makes this entry: "Aftor noon sattled about the size of the arch over the river Marsee [Mersey] to be 66 foot span and rise 16·4 feet." Next day he is "landing balk out of the ould river in to the canal." Then he goes on, "I prosceded to Worsley Mug was corking ye boats the masons woss making the senter of the waire [weir]. Whith[e] was osing to put the lator side of the water-wheel srouds on I orderd the pit for ye spindle of ye morter-mill to be sunk level with ye canal Mr. Gilbert sade ye 20 Tun Boat should be at ye water mitang [meeting] by 7 o'clock the next morn." Next morning he is on the works at Cornhill, setting "a carpentar to make scrwos" [screws], superintending the gravelling of the towing-path, and arranging with a farmer as to Mr. Gilbert's slack. And so he goes on from day to day with the minutest details of the undertaking.

He was not without his petty "werrets" and troubles either. Brindley and Gilbert do not seem to have got on very well together. They were both men of strong tempers, and neither would tolerate the other's interference. Gilbert, being the Duke's factotum, was accustomed to call Brindley's men from their work, which the other would not brook. Hence we have this entry on one occasion,—"A meshender [messenger] from Mr G I retorned the anser No more sosiety." In fact, they seem to have quarrelled.*

We find the following further entries on the subject in Brindley's note-book : "Thursday 17 Novr past 7 o'clock at night M Gilbert and sun Tom caled on mee at Gorshill and I went with them to ye Coik [sign of the Cock] tha stade all night and the had balk [blank ?] bill of parsill 18 Fryday November 7 morn I went to the Cock and Bruck-fast with Gilberts he in davred to imploye ye carpinters at Cornhill in making door and window frames for a Building in Castle field and shades for the mynors in Dito and other things I want them to Saill Moor Hee took upon him diriction of ye back drains and likwaise such Lands as be twixt the 2 hous and ceep uper side the large farme and was displesed with such raing as I had pointed out."

Those differences between Brindley and Gilbert

* The Earl of Bridgewater, in his rambling 'Letter to the Parisians,' above referred to, alleges that the quarrel originated in Gilbert's horse breaking into the field where Brindley's mare was grazing, and doing her such injury that the engineer was for a time prevented using the animal in the pursuit of his business. The mare was a great favourite of Brindley's, and he is said to have taken the thing very much to heart. The Earl alleges that Brindley was under the impression Gilbert had contrived the trick out of spite.

were eventually reconciled, most probably by the
mediation of the Duke, for the services of both
were alike essential to him; and we afterwards
find them working cordially together and consulting each other as before on any important part of
the undertaking.

During the construction of Longford Bridge,
Brindley seems, from his note-book, to have entertained considerable apprehensions as to its ability to
resist the heavy floods with which it was threatened.

Longford Bridge.

Thus, on the 26th of November, 1763, he enters :—
"Grate Rains the canal rose 2 inches extra I dreed
fr [4 ?] clock at Longfoard;" and on the following
day, which was a Sunday, he writes :—" Lay in
Bad till noon floode and Raine." Then in the
afternoon he adds, "The water in Longfoord
Brook was withe in six inches of the high of the
santer [centre] of ye waire [weir ?]." The bridge,
however, stood firm; and when the flood subsided,
the building was again proceeded with; and by the
end of the year it was finished and gravelled over,

while the embankment was steadily proceeding beyond the Mersey in the manner above described.

Brindley did not want for good workmen to carry out his plans. He found plenty of labourers in the neighbourhood accustomed to hard work, who speedily became expert excavators; and though there was at first a lack of skilled carpenters, blacksmiths, and bricklayers, they soon became trained into such under the vigilant eye of so able a master as Brindley was. We find him, in his note-book, often referring to the men by their names, or rather byenames; for in Lancashire proper names seem to have been little used at that time. "Black David" was one of the foremen most employed on difficult matters, and "Bill o Toms" and "Busick Jack" seem also to have been confidential workmen in their respective departments. We are informed by a gentleman of the neighbourhood * that most of the labourers employed were of a superior class, and some of them were "wise" or "cunning men," blood-stoppers, herb-doctors, and planet-rulers, such as are still to be found in the neighbourhood of Manchester. Their very superstitions, says our informant, made them thinkers and calculators. The foreman bricklayer, for instance, as his son used afterwards to relate, always "ruled the planets to find out the lucky days on which to commence any important work," and he added, "none of our work ever gave way." The skilled men had their trade-secrets, in which the unskilled were duly initiated,—simple matters in themselves, but not without their uses. The following may be taken

* R. Rawlinson, Esq., C.E., Engineer to the Bridgewater Canal.

as specimens of the secrets of embanking in those days :—

A wet embankment can be prevented from slipping by dredging or dusting powdered lime in layers over the wet clay or earth.

Sand or gravel can be made water-tight by shaking it together with flat bars of iron run in some depth, say two feet, and washing down loam or soil as the bars are moved about, thus obviating the necessity for clay puddle.

Dry-rot can be prevented in warehouses by setting the bricks opposite the ends of the main beams of the warehouse in dry sand.

Whilst constructing the canal, Brindley was very intimate with one Lawrence Earnshaw, of Mottram-in-Longdendale, a kindred mechanical genius, though in a smaller way. Lawrence was a very poor man's son, and had served a seven years' apprenticeship to the trade of a tailor, after which he bound himself apprentice to a clothier for seven years; but these trades not suiting his tastes, and being of a decidedly mechanical turn, he finally bound himself apprentice to a clock-maker, whom he also served for seven years. This eccentric person invented many curious and ingenious machines, which were regarded as of great merit in his time. One of these was an astronomical and geographical machine, beauti-fully executed, showing the earth's diurnal and annual motion, after the manner of an orrery. The whole of the calculations were made by him-self, and the machine is said to have been so exactly contrived and executed that, provided the vibration of the pendulum did not vary, the machine would not alter a minute in a hundred years; but this might probably be an extravagant estimate on the part of Earnshaw's friends. He was also a musical instrument maker and music teacher, a worker in metals and in wood, a painter and glazier, an optician, a bellfounder, a chemist and metallurgist, an engraver—in short, an almost universal mechanical genius. But though he could make all these things, it is mentioned as a remarkable fact, that with all his ingenuity, and after many efforts (for he made many), he never could make a wicker-basket! Indeed, trying to be a universal genius was his ruin. He did, or attempted to do, so much, that he never stood still and established himself in any one thing; and, notwithstanding his great ability, he died "not worth a groat." Amongst Earnshaw's various contrivances was a piece of machinery to raise water from a coal-mine at Hague, near Mottram, and (about 1753) a machine to spin and reel cotton at one operation —in fact, a spinning-jenny—which he showed to some of his neighbours as a curiosity, but, after having convinced them of what might be done by its means, he immediately destroyed it, saying that "he would not be the means of taking bread out of the mouths of the poor."* He was a total abstainer from strong drink, long before the days of Teetotal Societies. Towards the end of his life he continued on intimate terms with Brindley; and when they met they did not readily separate.

* Mr. Bennet Woodcroft, of the Patent Office, writes us as follows with reference to Earnshaw's alleged invention of a spinning machine:—"The fact really is, that the machine in question was invented by John Kay of Bury; and when, in 1753, a mob broke into Kay's house, and completely gutted it, the model of the spinning machine was saved by Earnshaw, who subsequently destroyed it."

While the undertaking was in full progress, from four to six hundred men were employed; they were divided into gangs of about fifty, each of which was directed by a captain and setter-out of the works. One who visited the canal during its construction in 1765, wrote thus of the busy scene which the works presented: "I surveyed the Duke's men for two hours, and think the industry of bees or labour of ants is not to be compared to them. Each man's work seems to depend on and be connected with his neighbour's, and the whole posse appeared as I conceive did that of the Tyrians when they wanted houses to put their heads in at Carthage." * At Stretford the visitor found "four hundred men at work, putting the finishing stroke to about two hundred yards of the canal, which reached nearly to the Mersey, and which, on drawing up the floodgates, was to receive a proper quantity of water and a number of loaded barges. One of these appeared like the hull of a collier, with its deck all covered, after the manner of a cabin, and having an iron chimney in the centre; this, on inquiry, proved to be the carpentry, but was shut up, being Sabbath-day, as was another barge, which contained the smith's forge. Some vessels were loaded with soil, which was put into troughs (see cut, p. 248), fastened together, and rested on boards that lay across two barges; between each of these there was room enough to discharge the loading by loosening some iron pins at the bottom of the troughs. Other barges lay loaded with the foundation-stones of the canal bridge, which is to carry the navigation across the Mersey. Near two thousand

* 'St. James's Chronicle,' July 1st, 1765.

oak piles are already driven to strengthen the foundations of this bridge. The carpenters on the Lancashire side were preparing the centre frame, and on the Cheshire side all hands were at work in bringing down the soil and beating the ground adjacent to the foundations of the bridge, which is designed to be covered with stone in a month, and finished in about ten days more." *

By these vigorous measures the works proceeded rapidly towards completion. Before, however, they had made any progress at the Liverpool end, Earl Gower, encouraged and assisted by the Duke, had applied for and obtained an Act to enable a line of navigation to be formed between the Mersey and the Trent; the Duke agreeing with the promoters of the undertaking to vary the course of his canal and meet theirs about midway between Prestonbrook and Runcorn, from which point it was to be carried northward towards the Mersey, descending into that river by a flight of ten locks, the total fall being not less than 79 feet from the level of the canal to low-water of spring-tides.

When this deviation was proposed, the bold imagination of Brindley projected a bridge across the tideway of the Mersey itself, which was there some four hundred and sixty yards wide, with the object of carrying the Duke's navigation directly onward to the port of Liverpool on the Lancashire side of the river.† This was an admirable idea,

---

* 'A History of Inland Navigations. Particularly those of the Duke of Bridgewater in Lancashire and Cheshire.' 2nd Ed., p. 39.

† This bold scheme, so earnestly advocated by Brindley, was thus noticed by a Liverpool paper of the time :—" On Monday last, Mr. Brindley waited upon several of the principal gentlemen of

which, if carried out, would probably have redounded more to the fame of Brindley than any other of his works.   But the cost of that portion of the canal which had already been executed, had reached so excessive an amount, that the Duke was compelled to stop short at Runcorn, at which place a dock was constructed for the accommodation of the shipping employed in the trade connected with the undertaking.

From Runcorn, it was arranged that the boats should navigate by the open tideway of the Mersey

The Duke's Dock, Liverpool.

to the harbour of Liverpool, at which place the Duke made arrangements to provide another dock for their accommodation.   Brindley made frequent

this town and others at Runcorn, in order to ascertain the expense that may attend the building of a bridge over the river Mersey at the latter place, which is estimated at a sum inferior to the advantages that must arise both to the counties of Lancaster and Chester from a communication of this sort."—Williamson's ' Liverpool Advertiser,' July 19th, 1768.

visits to Liverpool for the purpose of directing its excavation, and he superintended it until its completion. The Duke's Dock lies between the Salthouse and Albert Docks on the north, and the Wapping and King's Docks on the south. The Salthouse was the only public dock near it at the time that Brindley excavated this basin. There were only three others in Liverpool to the north, and not one to the south; but the Duke's Dock is now the centre of about five miles of docks, extending from it on either side along the Lancashire shore of the Mersey; and it continues to this day to be devoted to the purposes of the navigation.

CHAPTER V.

THE DUKE'S DIFFICULTIES—COMPLETION OF THE
CANAL—GROWTH OF MANCHESTER.

LONG before the Runcorn locks were constructed,
and the canal from Longford Bridge to the Mersey
was available for purposes of traffic, the Duke found
himself reduced to the greatest straits for want
of money. Numerous unexpected difficulties had
occurred, so that the cost of the works considerably
exceeded his calculations ; and though the engineer
carried on the whole operations with the strictest
regard to economy, the expense was nevertheless
almost more than any single purse could bear.
The execution of the original canal from Worsley
to Manchester cost about a thousand guineas a
mile, besides the outlay upon the terminus at
Manchester. There was also the expenditure incurred in building the requisite boats for the canal,
in opening out the underground workings of the
collieries at Worsley, and in erecting various mills,
workshops, and warehouses for carrying on the
new business.

The Duke was enabled to do all this without
severely taxing his resources, and he even entertained the hope of being able to grapple with the
still greater undertaking of cutting the twenty-four
miles of new canal from Longford Bridge to the

Mersey. But before these works were half finished, and whilst the large amount of capital invested in them was lying entirely unproductive, he found that the difficulties of the undertaking were likely to prove too much for him. Indeed, it seemed an enterprise beyond the means of any private person, and more like that of a monarch with State revenues at his command, than of a young English nobleman with only his private resources.

But the Duke was possessed by a brave spirit. He had put his hand to the work, and he would not look back. He had become thoroughly inspired by his great idea, and determined to bend his whole energies to the task of carrying it out. He was only thirty years of age—the owner of several fine mansions in different parts of the country, surrounded by noble domains—he had a fortune sufficiently ample to enable him to command the pleasures and luxuries of life, so far as money can secure them; yet he voluntarily denied himself their enjoyment, and chose to devote his time to consultations with an unlettered engineer, and his whole resources to the cutting of a canal to unite Liverpool and Manchester.

Taking up his residence at the Old Hall at Worsley—a fine specimen of the old timbered houses so common in South Lancashire and the neighbouring counties,—he cut down every unnecessary personal expense; denied himself every superfluity, except perhaps that of a pipe of tobacco; paid off his retinue of servants; put down his carriages and town house; and confined himself and his Ducal establishment to a total expenditure of 400*l.* a-year. A horse was, however, a necessity, for the purpose of enabling him to visit the canal works during their progress at distant points; and he accordingly continued to maintain one horse for himself and another for his groom.

Notwithstanding this rigid economy, the Duke still found his resources inadequate to meet the heavy cost of vigorously carrying on the undertaking, and on Saturday nights he was often put to the greatest shifts to raise the requisite money to pay his large staff of craftsmen and labourers.

Worsley Old Hall.

Sometimes their payment had to be postponed for a week or more, until the cash could be raised by sending round for contributions among the Duke's tenantry. Indeed, his credit fell to the lowest ebb, and at one time he could not get a bill for 500*l.* cashed in either Liverpool or Manchester.*

---

* There is now to be seen at Worsley, in the hands of a private person, a promissory note given by the Duke, bearing interest, for as low a sum as five pounds. Amongst the persons known to be lenders of money, to whom the Duke applied at the time, was

He was under the necessity of postponing all payments that could be avoided, and it went abroad that the Duke was "drowned in debt." He tried to shirk even the payment of his tithes, and turned a deaf ear to all the applications of the collector. At length the rector himself determined to waylay him. But the Duke no sooner caught sight of him coming across his path than he bolted! The rector was not thus to be baulked. He followed—pursued —and fairly ran his debtor to earth in a saw-pit! The Duke was not a little amused at being hunted in such a style by his parson, and so soon as he found his breath, he promised payment, which shortly followed.

When Mr. George Rennie, the engineer, was engaged, in 1825, in making the revised survey of the Liverpool and Manchester Railway, he lunched one day at Worsley Hall with Mr. Bradshaw, manager of the Duke's property, then a very old man. He had been a contemporary of the Duke, and knew of the monetary straits to which his Grace had been reduced during the construction of the works. Whilst at table, Mr. Bradshaw pointed to a small whitewashed cottage on the Moss, about a mile and a half distant, and said that in that cottage, formerly a public-house, the Duke, Brindley, and Gilbert had spent many an evening discussing the prospects of the canal while in progress. One of the principal topics of conversation on those occasions was the means of raising funds against the next pay night. "One evening in particular," said Mr. Bradshaw, "the party was

Mr. C. Smith, a merchant at Rochdale : but he would not lend a farthing, believing the Duke to be engaged in a perfectly ruinous undertaking.

unusually dull and silent. The Duke's ready-money was exhausted; the canal was not nearly finished; his Grace's credit was at the lowest ebb; and he was at a loss what step to take next. There they sat, in the small parlour of the little public-house, smoking their pipes, with a pitcher of ale before them, melancholy and silent. At last the Duke broke the silence by asking in a querulous tone, 'Well, Brindley, what's to be done now? How are we to get at the money for finishing this canal?' Brindley, after a few long puffs, answered through the smoke, 'Well, Duke, I can't tell; I only know that if the money can be got, I can finish the canal, and that it will pay well.' 'Ay,' rejoined the Duke, 'but *where* are we to get the money?' Brindley could only repeat what he had already said; and thus the little party remained in moody silence for some time longer, when Brindley suddenly started up and said, 'Don't mind, Duke; don't be cast down; we are sure to succeed after all!' The party shortly after separated, the Duke going over to Worsley to bed, to revolve in his mind the best mode of raising money to complete his all-absorbing project."

One of the expedients adopted was to send Gilbert, the agent, upon a round of visits among the Duke's tenants, raising five pounds here and ten pounds there, until he had gathered together enough to pay the week's wages. Whilst travelling about among the farmers on one of such occasions, Gilbert was joined by a stranger horseman, who entered into conversation with him; and it very shortly turned upon the merits of their respective horses. The stranger offered to swap with Gilbert, who, thinking the other's horse better than his own, agreed to the exchange. On afterwards alighting at a lonely village inn, which he had not before frequented, Gilbert was surprised to be greeted by the landlord with mysterious marks of recognition, and still more so when he was asked if he had got a good booty. It turned out that he had exchanged horses with a highwayman, who had adopted this expedient for securing a nag less notorious than the one which he had exchanged with the Duke's agent.*

At length, when the tenantry could furnish no further advances, and loans were not to be had on any terms in Manchester or Liverpool, and the works must needs come to a complete stand unless money could be raised to pay the workmen, the Duke took the road to London on horseback, attended only by his groom, to try what could be done with his London bankers. The house of Messrs. Child and Co., Temple Bar, was then the principal banking-house in the metropolis, as it is the oldest ; and most of the aristocratic families kept their accounts there. The Duke had determined at the outset of his undertaking not to mortgage his landed property, and he had held to this resolution. But the time arrived when he could not avoid borrowing money of his bankers on such other security as he could offer them. He had already created a valuable and lucrative property, which was happily available for the purpose. The canal from Worsley to Manchester had proved remunerative in an extraordinary degree, and was already producing a large income. He had not the same scruples as to the pledging of the revenues

* The Earl of Ellesmere's 'Essays on History, Biography,' &c., p. 236.

of his canal that he had to the mortgaging of his lands; and an arrangement was concluded with the Messrs. Child under which they agreed to advance the Duke sums of money from time to time, by means of which he was eventually enabled to finish the entire canal.

The Messrs. Child and Co. have kindly permitted an examination of their books to be made for the purposes of this memoir; and we are accordingly enabled to state that from them it appears that the Duke obtained his first advance of 3,800*l.* from the firm about the middle of the year 1765, at which time he was in the greatest difficulty; shortly after a further sum of 15,000*l.*; then 2,000*l.*, and various other sums, making a total of 25,000*l.*; which remained owing until the year 1769, when the whole was paid off—doubtless from the profits of the canal traffic as well as the economised rental of the Duke's unburthened estates.

The entire level length of the new canal from Longford Bridge to the upper part of Runcorn, nearly twenty-eight miles in extent, was finished and opened for traffic in the year 1767, after the lapse of about five years from the passing of the Act. The formidable flight of locks, from the level part of the canal down to the waters of the Mersey at Runcorn, were not finished for several years later, by which time the receipts derived by the Duke from the sale of his coals and the local traffic of the undertaking enabled him to complete them with comparatively little difficulty. Considerable delay was occasioned by the resistance of an obstinate landowner near Runcorn, Sir Richard Brooke, who interposed every obstacle which it was in his power to offer; but his opposition too was at length overcome, and the new and complete line of water-communication between Manchester and Liverpool was finally opened throug h-out.

In a let-ter written from Run-corn, dated the 1st January, 1773, we find it stated that "yesterday the locks were opened, and the ' Heart of Oak,' a vessel of 50 tons burden, for Liver-pool, passed

The Locks at Runcorn.

through them. This day, upwards of six hundred of his Grace's workmen were entertained upon the lock banks with an ox roasted whole and plenty of good liquor. The Duke's health and many other toasts were drunk with the loudest acclamations by the multitude, who crowded from all parts of the country to be spectators of these astonishing works. The gentlemen of the country for a long time entertained a very unfavourable opinion of this undertaking, esteeming it too difficult to be accomplished, and fearing their lands would be cut and defaced without producing any real benefit to themselves or the public; but they now see with pleasure that their fears and apprehensions were ill-grounded, and they join with one voice in applauding the work, which cannot fail to produce the most beneficial consequences to the landed property, as well as to the trade and commerce of this part of the kingdom."

Whilst the canal works had been in progress, great changes had taken place at Worsley. The Duke had year by year been extending the workings of the coal; and when the King of Denmark, travelling under the title of Prince Travindahl, visited the Duke in 1768, the tunnels had already been extended for nearly two miles under the hill. When the Duke began the works, he possessed only such of the coal-mines as belonged to the Worsley estate; but he purchased by degrees the adjoining lands containing seams of coal which run under the high ground between Worsley, Bolton, and Bury; and in course of time the underground canals connecting the different workings extended for a distance of nearly forty miles. Both the hereditary and the purchased mines are worked upon two main levels, though in all there are four different levels, the highest being a hundred and twenty yards above the lowest. In opening up the underground workings the Duke is said to have expended about 168,000*l*.; but the immense revenue derived from the sale of the coals by canal rendered this an exceedingly productive outlay. Besides the extension of the canal along these tunnels, the Duke subsequently carried a branch by the edge of Chat-Moss to Leigh, by which means new supplies of coal were introduced to Manchester from that district, and the traffic was still further increased. It was a saying of the Duke's, that "a navigation should always have coals at the heels of it."

The total cost of completing the canal from Worsley to Manchester, and from Longford Bridge to the Mersey at Runcorn, amounted to 220,000*l*. A truly magnificent undertaking, nobly planned and nobly executed. The power imparted by riches was probably never more munificently exercised than in this case; for, though the traffic proved a source of immense wealth to the Duke, it also conferred incalculable blessings upon the population of the district. It added much to their comforts, increased their employment, and facilitated the operations of industry in all ways. As soon as the canal was opened its advantages began to be felt. The charge for water-carriage between Liverpool and Manchester was lowered one-half. All sorts of produce were brought to the latter town, at moderate rates, from the farms and gardens adjacent to the navigation, whilst the value of agricultural property was immediately raised by the facilities afforded for the conveyance of lime and manure, as well as by reason of the more ready

access to good markets which it provided for the
farming classes. The Earl of Ellesmere has not
less truly than elegantly observed, that "the history
of Francis Duke of Bridgewater is engraved in
intaglio on the face of the country he helped to
civilise and enrich."

Probably the most remarkable circumstance
connected with the money history of the enterprise is this : that although the canal yielded an
income which eventually reached about 80,000*l*. a
year, it was planned and executed by Brindley at
a rate of pay considerably less than that of an
ordinary mechanic of the present day. The highest
wage he received whilst in the employment of the
Duke was 3*s*. 6*d*. a day. For the greater part of
the time he received only half-a-crown. Brindley,
no doubt, accommodated himself to the Duke's
pinched means, and the satisfactory completion of
the canal was with him as much a matter of disin-
terested ambition and of professional character as
of pay. He seems to have kept his own expenses
down to the very lowest point. Whilst superintending the works at Longford Bridge, we find
him making an entry of his day's personal charges
at only 6*d*. for "ating and drink." On other days
his outgoings were confined to "2*d*. for the turnpike." When living at 'The Bull,' near the works
at Throstle Nest, we find his dinner costing 8*d*.
and his breakfast 6*d*. His expenditure throughout
was on an equally low scale, for he studied in all
ways to economise the Duke's means, that every
available shilling might be devoted to the prosecution of the works.

The Earl of Bridgewater, in his singular publication, the 'Letter to the Parisians,' above referred

I.   T

to, states that "Brindley offered to stay entirely
with the Duke, and do business for no one else,
if he would give him a guinea a week;" and this
statement is repeated by the late Earl of Ellesmere
in his 'Essays on History, Biography,' &c. But,
on the face of it, the statement looks untrue; and
we have since found, from Brindley's own note-
book, that on the 25th of May, 1762, he was receiving a guinea a day from the Earl of Warrington
for performing services for that nobleman; nor is
it at all likely that he would prefer the Duke's
three-and-sixpence a day to the more adequate rate
of payment which he was accustomed to charge and
to receive from other employers. It is quite true,
however—and the fact is confirmed by Brindley's
own record — that he received no more than a
guinea a week whilst in the Duke's service; which
only affords an illustration of the fact that eminent
constructive genius may be displayed and engineering greatness achieved in the absence of any
adequate material reward.

In a statement of the claims of Brindley's representatives, forwarded to the Earl of Bridgewater
on the 3rd of November, 1803, it was stated that
"during the period of his employ under His Grace,
many highly advantageous and lucrative offers were
made to him, particularly one from the Prince of
Hesse, in 1766, who at that time was meditating
a canal through his dominions in Germany, and
who offered to subscribe to any terms Mr. Brindley
might stipulate. To this engagement his family
strongly urged him, but the solicitation of the
Duke, in this as in every other instance, to remain
with him, outweighed all pecuniary considerations; relying upon such a remuneration from His

Grace as the profits of his work might afterwards justify."*

The inadequate character of his remuneration was doubtless well enough known to Brindley himself, and rendered him very independent in his bearing towards the Duke. They had frequent

* We regret to have to add that Brindley's widow (afterwards the wife of Mr. Williamson, of Longport) in vain petitioned the Duke and his representatives, as well as the above Earl of Bridgewater, for payment of a balance said to have been due to Brindley for services, at the time of the engineer's death. In her letter to Robert Bradshaw, M.P., dated the 2nd May, 1803, Mrs. Williamson says : "It will doubtless appear to you extraordinary that so very late an application should now be made . . . but I must beg leave to state that repeated applications were made by me (after Mr. Brindley's sudden and unexpected death) to the late Mr. Thomas Gilbert and also to his brother, but without any other effect than that of constant promises to lay the matter before His Grace; and I conceive it owing to this channel of application that no settling ever took place. A letter was also written to His Grace on this subject so late as the year 1801, but no answer was received. From the year 1765 to 1772, Mr. Brindley received no money on account of his salary. At that time he was frequently in very great want, and made application to the Duke, whose answer (to use the Duke's expression) was, 'I am much more distressed for money than you; however, as soon as I can recover myself, your services shall not go unrewarded.' In consequence of this, Mr. Brindley was under the necessity of borrowing several sums to make good engagements he was then under to various canal companies. In the year 1774, two years after Mr. Brindley's death, the late Mr. John Gilbert paid my brother, Mr. Henshall, the trifling sum of 100l. on account of Mr. Brindley's time, which is all that has been received. I beg leave to suggest how small and inadequate a return this is for his services during a period of seven years. Mr. B.'s travelling expenses on His Grace's account during that time were considerable, towards which, when he had not sufficient money to carry him the whole journey, he now and then received a small sum. How far his plans and undertakings have been beneficial to His Grace's interest is well known."

differences as to the proper mode of carrying on
the works; but Brindley was quite as obstinate as
the Duke on such occasions, and when he felt convinced that his own plan was the right one he
would not yield an inch. It is said that, after long
evening discussions at the hearth of the old
timbered hall at Worsley, or at the Duke's house
at Liverpool, while the works there were in progress, the two would often part at night almost at
daggers-drawn. But next morning, on meeting
at breakfast, the Duke would very frankly say to
his engineer, " Well, Brindley, I have been thinking over what we were talking about last night.
I find you may be right after all ; so just finish the
work in your own way."

The Duke himself, to the end of his life, took the
greatest personal interest in the working of his
coal-mines, his canals, his mills, and his various
branches of industry.   These were his hobbies,
and he took pleasure in nothing else. He was
utterly lost to the fashionable world, and, as some
thought, to a sense of its proprieties. Shortly after
his canal had been opened for the conveyance of
coals, the Duke established a service of passage-
boats between Manchester and Worsley, and
between Manchester and a station within two
miles of Warrington, by which passengers were
conveyed at the rate of a penny a mile.* The boats
were fitted up like the Dutch treckschuyts, and,

---

* When the Duke had put on the boats and established the
service, he offered to let them for 60l. a year ; but not being
able to find any person to take them at that price, he was under
the necessity of conducting the service himself, by means of an
agent. In the course of a short time the boats were found to yield
a clear profit of 1600l. a year.

being found cheap as well as convenient, were largely patronized by the public. This service was afterwards extended to Runcorn, and from thence to Liverpool.

The Duke took particular pleasure in travelling by his own boats, preferring them to any more stately and aristocratic method. He often went by them to Manchester to watch how the coal-trade was going on. When the passengers alighted at the coal-wharf, there were usually many poor people about, wheeling away their barrow-loads of coals. One of the Duke's regulations was, that whenever any deficiency in the supply was apprehended, those people who came with their wheelbarrows, baskets, and aprons for small quantities, should be served first, and waggons, carts, and horses sent away until the supply was more abundant. The numbers of small customers who thus resorted to the Duke's coal-yard rendered it a somewhat busy scene, and the Duke liked to look on and watch the proceedings.

One day a customer of the poorer sort, having got his sack filled, looked about for some one to help it on to his back. He observed a stoutish man standing near, dressed in a spencer, with dark drab smallclothes. "Heigh! mester!" said the man, "come, gie me a lift wi' this sack o' coal on to my shouder." Without any hesitation, the person in the spencer gave the man the required "lift," and off he trudged with the load. Some one near, who had witnessed the transaction, ran up to the man and asked, "Dun yo know who's that yo've been speaking tull?" "Naw! who is he?" "Why, it's th' Duke his-sen!" "The Duke!" exclaimed the man, dropping the bag of coals from his shoulder,

"Hey! what'll he do *at* me? Maun a goo an ax his pardon?" But the Duke had disappeared.*

He was very fond of watching his men at work, especially when any new enterprise was on foot. When they were boring for coal at Worsley, the Duke came every morning and looked on for a long time together. The men did not like to leave off work whilst he remained there, and they became so dissatisfied at having to work so long beyond the hour at which the bell rang, that Brindley had difficulty in getting a sufficient number of hands to continue the boring. On inquiry, he found out the cause and communicated it to the Duke, who from that time made a point of immediately walking off when the bell rang, returning when the men had resumed work, and remaining with them usually until six o'clock. He observed, however, that though the men dropped work promptly as the bell rang, when he was not by, they were not nearly so punctual in resuming work, some straggling in many minutes after time. He asked to know the reason, and the men's excuse was, that though they could always hear the clock when it struck twelve, they could not so readily hear it when it struck only one. On this, the Duke had the mechanism of the clock altered so as to make it strike *thirteen* at one o'clock; which it continues to do to this day.

* A similar story is told of him at Worsley. A boy had been fetching some coal from the mouth of the tunnel, and having rested his load could not get it cleverly on his back again. Seeing but not knowing the Duke, he called out, " Here, felly, gie us a hoist up." The Duke asked the boy a number of questions before helping him up with the sack ; but when the youth felt it safe for carrying, he expressed his acknowledgments by observing " Wal. thou's a big chap, but thou's a lazy un ! "

On another occasion, going into the yard at
Worsley, he saw two men employed in grinding
an axe, and three others looking on, probably
waiting their turn at the grindstone. The Duke
said nothing; but next morning he was in the yard
early, and said to the foreman that he had observed
it took five men to grind an axe. He then ordered
that a water-wheel should be put up to drive the
grindstone, and it was set about at once. The
Duke was often after seen grinding the ferrule of
his walking-stick against the self-acting machine.

His time was very fully occupied with his
various business concerns, to which he gave a
great deal of personal attention. Habit made him
a business man—punctual in his appointments,
precise in his arrangements, and economical both
of money and time. When it was necessary for
him to see any persons about matters of business,
he preferred going to them instead of letting them
come to him; "for," said he, "if they come to me,
they may stay as long as they please; if I go to
them, I stay as long as *I* please." His enforced
habits of economy during the construction of the
canal had fully impressed upon his mind the value
of money. Yet, though "near," he was not penu-
rious, but was usually liberal, and sometimes muni-
ficent. When the Loyalty Loan was raised, he
contributed to it no less a sum than 100,000*l*. in
cash. He was thoroughly and strongly national,
and a generous patron of many public benevolent
institutions.

The employer of a vast number of workpeople,
he exercised his influence over them in such a
manner as to evoke their gratitude and blessings.
He did not "lord it" over them, but practically

taught them, above all things, to help themselves. He was the pattern employer of his neighbour-hood. With a kind concern for the welfare of his colliery workmen—then a half-savage class—he built comfortable dwellings and established shops and markets for them ; by which he ensured that at least a certain portion of their weekly earnings should go to their wives and families in housing, food, and clothing, instead of being squandered in idle dissipation and drunkenness.

In order to put a stop to idle Mondays, he im-posed a fine of half-a-crown on any workman who did not go down the pit at the usual hour on that morning ; and hence the origin of what is called Half Crown Row at Worsley, as thus described by one of the colliers :—" T'ould dook fined ony mon as didn't go daown pit o' Moonday mornin auve a craown, and abeaut thot toime he made a new road to t'pit, so t'colliers caw'd it Auve Craown Row."

Debts contracted by the men at public-houses were not recognised by the pay-agents. The steadi-est workmen were allowed to occupy the best and pleasantest houses as a reward for their good con-duct. The Duke also bound the men to contribute so much of their weekly earnings to a general sick club ; and he encouraged a religious tone of cha-racter amongst his people by the establishment of Sunday schools, which were directly superintended by his agents, selected from the best available class. The consequence was, that the Duke's colliers soon held a higher character for sobriety, intelligence, and good conduct, than the weavers and other workpeople of the adjacent country.

He did not often visit London, where he had

long ceased to maintain a house; but when he went there he made an arrangement with one of his friends, who undertook for a stipulated sum to provide a daily dinner for His Grace and a certain number of guests whilst he remained in town. He also made occasional visits to his fine estate of Ashridge, in Buckinghamshire, taking the opportunity of spending a few days, going or coming, with Earl Gower and his Countess, the Duke's only sister, Lady Louisa Egerton, at Trentham Park. During his visits at the latter place, the Duke would get ensconced on a sofa in some distant corner of the room in the evenings, and discourse earnestly to those who would listen to him about the extraordinary advantages of canals. There was a good deal of fun made on these occasions about "the Duke's hobby." But he was always like a fish out of water until he got back to Worsley, to John Gilbert, his coal-pits, his drainage, his mills, and his canals.

No wonder he was fond of Worsley. It had been the scene of his triumphs, and the foundation of his greatness. Illustrious visitors from all parts resorted thither to witness Brindley's "castle in the air," and to explore the underground canals at Worsley-hill. Frisi, the Italian, the King of Denmark, and others, regarded these subterranean works with wonder and admiration when they were only from $1\frac{1}{2}$ to 2 miles in length; soon they extended to nearly 40 miles. Among the visitors entertained by the Duke was Fulton, the American artist, with whose speculations he was much interested. Fulton had given his attention to the subject of canals, and was then speculating on the employment of steam power for propelling canal boats. The Duke was so much impressed with Fulton's ingenuity, that he urged him to give up the profession of a painter and devote himself to that of a civil engineer. Fulton acted on his advice, and shortly after we find him residing at Birmingham—the central workshop of England—studying practical mechanics, and fitting himself for superintending the construction of canals, on which he was afterwards employed in the midland counties.*

The Duke did not forget the idea which Fulton had communicated to him as to the employment of steam as a motive power for boats, instead of horses; and when he afterwards heard that Symington's steam-boat, 'The Dundas,' had been tried successfully on the Forth and Clyde Canal, he arranged to have six canal boats constructed after Symington's model; for he was a man to shrink from no expense in carrying out an enterprise which, to use his own words, had "utility at the heels of it." The Earl of Ellesmere, in his 'Essay on Aqueducts and Canals,' states that the Duke made actual experiment of a steam-tug, and quotes the following from the communication of one of the Duke's servants, alive in 1844: "I well remember the steam-tug experiment on the canal. It was between 1796 and 1799. Captain Shanks, R.N., from Deptford, was at Worsley many weeks preparing it, by the Duke's own orders and under his own eye. It

* The treatise which Fulton afterwards published, entitled 'A Treatise on Canal Navigation, exhibiting the numerous advantages to be derived from small Canals, &c., with a description of the machinery for facilitating conveyance by water through the most mountainous countries, independent of Locks and Aqueducts,' (London, 1796,) is well known amongst engineers.

was set going and tried with coal-boats ; but it
went slowly, and the paddles made sad work with
the bottom of the canal, and also threw the water
on the bank. The Worsley people called it Bona-
parte."* But the Duke dying shortly after, the
trustees refused to proceed with the experiment,
and the project consequently fell through. Had
the Duke lived, canal steam-tugs would doubtless
have been fairly tried; and he might thus have
initiated the practical introduction of steam-navi-
gation in England, as he unquestionably laid the
foundations of the canal system. He lived long
enough, however, to witness the introduction of
tram-roads, and he saw considerable grounds for
apprehension in them. "We may do very well,"
he once observed to Lord Kenyon, "if we can keep
clear of these —— tram-roads."

He was an admirable judge of character, and
was rarely deceived as to the men he placed confidence in. John Gilbert was throughout his confidential adviser—a practical out-doors man, full
of energy and perseverance. When any proposal
was made to the Duke, he would say, "Well, thou
must go to Gilbert and tell him all about it ; I'll
do nothing without I consult him." From living
so much amongst his people, he had contracted
their style of speaking, and "thee'd" and "thou'd"
those whom he addressed, after the custom of the
district. He was rough in his speech, and gruff and
emphatic in his manner, like those amidst whom he
lived ; but with the rough word he meant and did
the kindly act. His early want of education de-
barred him in a measure from the refining influences
of letters ; for he read little, except perhaps an

* Lord Ellesmere's 'Essays,' p. 241.

occasional newspaper, and he avoided writing whenever he could. He also denied himself the graces
of female society; and the seclusion which his early
disappointment in love had first driven him to, at
length grew into a habit. He lived wifeless and
died childless. He would not even allow a woman
servant to wait upon him.

In person he was large and corpulent; and the
slim youth on whom the bet had been laid that he
would be blown off his horse when riding the race
in Trentham Park so many years before, had grown
into a bulky and unwieldy man. His features
strikingly resembled those of George III. and other
members of the Royal Family. He dressed care-
lessly, and usually wore a suit of brown—something of the cut of Dr. Johnson's—with dark drab
breeches, fastened at the knee with silver buckles.
At dinner he rejected, with a kind of antipathy, all
poultry, veal, and such like, calling them "white
meats," and wondered that everybody, like himself,
did not prefer the brown. He was a great smoker,
and smoked far more than he talked. Smoking was
his principal evening's occupation when Brindley
and Gilbert were pondering with him over the
difficulty of raising funds to complete the navigation, and the Duke continued his solitary enjoyment through life. One of the droll habits to which
he was addicted was that of rushing out of the
room every five minutes, with the pipe in his
mouth, to look at the barometer. Out of doors
he snuffed, and he would pull huge pinches out of
his right waistcoat pocket and thrust the powder up
his nose, accompanying the operation with sundry
strong short snorts.

He would have neither conservatory, pinery,

flower-garden, nor shrubbery at Worsley ; and once, on his return from London, finding some flowers which had been planted in his absence, he

Francis Duke of Bridgewater (æt. 70.)

whipped their heads off with his cane, and ordered them to be rooted up. The only new things introduced about the place were some Turkey oaks, with which his character seemed to have more sympathy. But he took a sudden fancy for pictures, and with his almost boundless means the formation of a valuable collection of pictures was easy.*

Lord Ellesmere says : "An accident laid the foundation of the Bridgewater collection. Dining one day with his nephew, Lord Gower, afterwards Duke of Sutherland, the Duke saw and admired a picture which the latter had picked up a bargain, for some 10*l.*, at a broker's in the morning. 'You must take me,' he said, 'to that —— fellow tomorrow.' Whether this impetuosity produced any immediate result we are not informed, but plenty of such 'fellows' were doubtless not wanting to cater for the taste thus suddenly developed."

Fortunately the Duke's investments in paintings appear to have been well directed; and a discerning eye seems to have guided a liberal hand in selecting fine separate works, as well as the gems from Continental collections which were then dispersed and found their way hither, thus enabling him to lay the foundation of the famous Bridgewater Gallery, one of the finest private collections in Europe. At his death, in 1803, its value was estimated at 150,000*l.*

The Duke very seldom took part in politics, but usually followed the lead of his relative Earl Gower, afterwards Marquis of Stafford, who was a Whig. In 1762, we find his name in a division on a motion to withdraw the British troops from

---

* "His purchases from Italy and Holland were judicious and important, and, finally, the distractions of France forcing the treasures of the Orleans Gallery into this country, he became a principal in the fortunate speculation of its purchase."—'Essays on History, Biography,' &c.

Germany, and on the loss of the motion he joined in a protest on the subject. When the repeal of the American Stamp Act was under discussion His Grace was found in the ranks of the opposition to the measure. He strongly supported Mr. Fox's India Bill, and generally approved the policy of that statesman.

The title of Duke of Bridgewater died with him. The Earldom went to his cousin General Egerton, seventh Earl of Bridgewater, and from him to his brother the crazed Francis Henry, eighth Earl; and on his death at Paris, in February, 1829, that title too became extinct. The Duke bequeathed about 600,000*l.* in legacies to his relatives, General Egerton, the Countess of Carlisle, Lady Anne Vernon, and Lady Louisa Macdonald. He devised most of his houses, his pictures, and his canals, to his nephew George Granville (son of Earl Gower), second Marquis of Stafford and first Duke of Sutherland, with reversion to his second son, Lord Francis Egerton, first Earl of Ellesmere, who thus succeeded to the principal part of the vast property created by the Duke of Bridgewater. The Duke was buried in the family vault at Little Gaddesden, Hertfordshire, in the plainest manner, without any state, at his own express request. On his monument was inscribed the simple and appropriate epitaph *Impulit ille rates ubi duxit aratra Colonus.*

The Duke was a great public benefactor. The boldness of his enterprise, and the salutary results which flowed from its execution, entitle him to be regarded as one of the most useful men of his age. A Liverpool letter of 1765 says, " The services the Duke has rendered to the town and neighbourhood of Manchester have endeared him to the country,

more especially to the poor, who, with grateful
benedictions, repay their noble benefactor."* If
he became rich through his enterprise, the public
grew rich with him and by him; for his undertaking was no less productive to his neighbours
than it was to himself. His memory was long
venerated by the people amongst whom he lived,
—a self-reliant, self-asserting race, proud of their
independence, full of persevering energy, and strong
in their attachments. The Duke was a man very
much after their own hearts, and a good deal after
their own manners. In respecting him, they were
perhaps but paying homage to those qualities
which they most cherished in themselves. Long
after the Duke had gone from amongst them, they
spoke to each other of his rough words and his
kindly acts, his business zeal and his indomitable
courage. He was the first great "Manchester man."
His example deeply penetrated the Lancashire
character, and his presence seems even yet to hover
about the district. "The Duke's canal" still carries
a large proportion of the merchandise of Manchester
and the neighbouring towns; "the Duke's horses"†
still draw "the Duke's boats;" "the Duke's coals"

* 'History of Inland Navigation,' p. 76.
† The Duke at first employed mules in hauling the canal-boats,
because of the greater endurance and freedom from disease of
those animals, and also because they could eat almost any description of provender. The Duke's breed of mules was for a long time
the finest that had been known in England. The popular impression in Manchester is, that the Duke's Acts of Parliament
authorising the construction of his canals, forbade the use of
horses, in order that men might be employed ; and that the Duke
consequently dodged the provisions of the Acts by employing
mules. But this is not the case, there being no clause in any of
them prohibiting the use of horses.

still issue from " the Duke's levels ; " and when any question affecting the traffic of the district is under consideration, the questions are still asked of "What will the Duke say ? "   " What will the Duke do ? " *

Manchester men of this day may possibly be surprised to learn that they owe so much to a Duke, or that the old blood has helped the new so materially in the development of England's modern industry. But it is nevertheless true that the Duke of Bridgewater, more than any other single man, contributed to lay the foundations of the prosperity of Manchester, Liverpool, and the surrounding districts. The cutting of the canal from Worsley to Manchester conferred upon that town the immediate benefit of a cheap and abundant supply of coal ; and when Watt's steam-engine became the great motive power in manufactures, such supply became absolutely essential to its existence as a manufacturing town. Being the first to secure this great advantage, Manchester thus got the start forward which she has never since lost. †

But, besides being a waterway for coal, the Duke's canal, when opened out to Liverpool, immediately conferred upon Manchester the immense advantage of direct connection with an

---

* There is even a tradition surviving at Worsley, that "the Duke" rides through the village once in every year at midnight, drawn by six coal-black horses !

† The cotton trade was not of much importance at first, though it rapidly increased when the steam-engine and spinning-jenny had become generally adopted. It may be interesting to know that sixty years since it was considered satisfactory if *one* cotton-flat a day reached Manchester from Liverpool. In the Duke's time the flats always "cast anchor" on their way, or at least laid up for the night, at six o'clock precisely, starting again at six o'clock on the following morning.

I   U

excellent seaport. New canals, supported by the
Duke and constructed by the Duke's engineer,
grew out of the original scheme between Manchester and Runcorn, which had the further effect
of placing the former town in direct water-communication with the rich districts of the north-
west of England. Then the Duke's canal terminus
became so important, that most of the new navigations were laid out to join it; those of Leigh,
Bolton, Stockport, Rochdale, and the West Riding
of Yorkshire, being all connected with the Duke's
system, whose centre was at Manchester. And thus
the whole industry of these districts was brought,
as it were, to the very doors of that town.

But Liverpool was not less directly benefited
by the Duke's enterprise. Before his canal was
constructed, the small quantity of Manchester
woollens and cottons manufactured for exportation was carried on horses' backs to Bewdley
and Bridgenorth on the Severn, from whence they
were floated down that river to Bristol, then the
chief seaport on the west coast. No sooner, however, was the new water-way opened out than the
Bridgenorth pack-horses were taken off, and the
whole export trade of the district was concentrated at Liverpool. The additional accommodation required for the increased business of the
port was promptly provided as occasion required.
New harbours and docks were built, and before
many years had passed Liverpool had shot far
ahead of Bristol, and became the chief port on the
west coast, if not in all England. Had Bristol
been blessed with a Duke of Bridgewater, the
result might have been altogether different; and
the valleys of Wilts, the coal and iron fields of

Wales, and the estuary of the Severn, might have been what South Lancashire and the Mersey are now.   Were statues any proof of merit, the Duke would long since have had the highest statue in Manchester as well as Liverpool erected to his memory, and that of Brindley would have been found standing by his side; for they were both heroes of industry and of peace, though even in commercial towns men of war are sometimes more honoured.

We can only briefly glance at the extraordinary growth of Manchester since the formation of the Duke's canal, as indicated by the annexed plan.

Though Manchester was a place of some importance about the middle of last century, it was altogether insignificant in extent, trade, and population, compared with what it is now.   It consisted of a few principal streets—narrow, dark, and tortuous—one of them leading from the Market Place to St. Ann's Square, being very appropriately named "Dark Entry."   Deansgate was the principal original street of the town, and so called because of its leading to the dean or valley along which it partly extended.   From thence a few streets diverged in different directions into the open country.   St. Ann's Square, the fashionable centre of modern Manchester, was in 1770 a corn-field surrounded with lofty trees, and known by the name of "Acre's Field."   The cattle-fairs of the town were held there, the entrance from Deansgate being by Toll Lane, a narrow, dirty, unpaved way, so called because toll was there levied on the cattle proceeding towards the fair.   The ancient seat of the Radcliffe family still stood at Pool Fold, close to the site of the

modern Cross Street, and the water in the moat
was used as a ducking-pond for scolds. When
the pool became filled up, the ducking-pond was
removed to Daub Holes, then on the outskirts of

Plan of Manchester, showing its extent at three periods.

(The parts printed black ⊟ represent Manchester in 1770 ; those dark-shaded ⬛ show
its extent in 1804 ; and the light-shaded parts ⬜ Manchester at the present day.)

the town, where the Infirmary now stands. The
site of King Street, now the very heart of Manchester, was as yet comparatively retired, a colony
of rooks having established themselves in the tall

trees at its upper end, from which they were only
finally expelled about forty years ago. Cannon
Street was the principal place of business, the
merchants and their families living in the comparatively humble tenements fronting the street,
the equally humble warehouses in which their
business was done standing in the rear. The
ground on which the crowded thoroughfares of
Oldham Street, London Road, Mosley Street, and
their continuations, now exist, was as yet but
garden or pasture-land. Salford itself was only
a hamlet occupying the bend of the Irwell. It
consisted of a double line of mean houses, extending from the Old Bridge (now Victoria Bridge)
to about the end of Gravel Lane, then a country
road containing only a few detached cottages.
The comparatively rural character of Manchester
may be inferred from the circumstance that the
Medlock and the Irk, the Tib and Shooter's
Brook, were favourite fishing streams. Salmon
were caught in the Medlock and at the mouth
of the Irk; and the others were well stocked with
trout. The Medlock and the Irk are now as black
as old ink, and as thick; but the Tib and Shooter's
Brook are entirely lost,—having been absorbed,
like the London Fleet, in the sewage system of
the town. Tib Street and Tib Lane indicate the
former course of the Tib; but of Shooter's Brook
not a trace is left.

The townships of Ardwick Green, Hulme, and
Chorlton-upon-Medlock (formerly called Chorlton
Row), were entirely rural. The old rate-books of
Chorlton Row exhibit some curious facts as to the
transformations effected in that township. In 1720,
a "lay" of 14_d._ in the pound produced a sum of

26*l*. 18*s*., the whole disbursements for the year amounting to 28*l*. 8*s*. 5*d*. From the highway rate laid in 1722, it appears that the contributors were only twenty persons in all, whose payments ranged from 8*d*. to 1*l*. 13*s*. 4d., producing a total levy of 6*l*. 18*s*. 10*d*. for the year. From the disbursements, it appears that the regular wage paid to the workmen employed was a shilling a-day. In 1750, a lay of 3*d*. in the pound produced only 6*l*. 2*s*. 1½*d*.; so that the population and value of property in Chorlton Row had not much increased during the thirty years that had passed. In 1770, two levies brought in 57*l*. 8*s*. 6*d*.; and in 1794, four, made in that year, produced 208*l*. 2*s*. 4*d*.* Among the list of contributors in the latter year we find "Mrs. Quincey 16*s*. 6*d*."—the mother of De Quincey, the English opium-eater, who was brought up in Chorlton Row. De Quincey describes the home of his childhood as a solitary house, "beyond which was nothing but a cluster of cottages, composing the little hamlet of Greenhill." It was connected by a winding lane with the Rusholme road. The house, called Greenheys — the nucleus of an immense suburban district—built by De Quincey's father, "was then," he says, "a clear mile from the outskirts of Manchester," Princess Street being then the termination of the town on that side.†
Now it is enveloped by buildings in all directions, and nothing of the former rural character of the

---

* Recent "poor-lays" exhibit a very different result from what they did in former years. In 1860-1 the poor-rate levied in Chorlton-upon-Medlock yielded (at 2*s*. 10*d*. in the pound) 18,798*l*.; the property in the township being of the rateable value of 145,844*l*.

† De Quincey's 'Autobiographic Sketches,' pp. 34, 48.

neighbourhood remains but the names of Greenhill, Rusholme, and Greenheys.

Coming down to the second expansion of Manchester, as exhibited on our plan, it will be observed that a considerable increase of buildings had taken place in the interval between 1770 and 1804. The greater part of the town was then contained in the area bounded by Deansgate, the crooked lanes leading to Princess Street, Bond Street, and David Street, to the Rochdale Canal, and round by Ancoats Lane (now Great Ancoats Street) and Swan Street, to Long Millgate, then a steep narrow lane forming the great highway into North Lancashire. Very few buildings existed outside the irregular quadrangle indicated by the streets we have named. The straggling houses of Deansgate, which were principally of timber, ended at Knott Mill. A few dye-works stood at intervals along the Medlock, now densely occupied by buildings for miles along both banks. Salford had not yet extended to St. Stephen's Street in one direction, nor above halfway to Broughton Bridge in another.* The comparatively limited spaces thus indicated sufficed, however, for places of business and habitations for the population. Now the central districts are almost exclusively occupied for business purposes, and houses for dwellings have rapidly extended in all directions. The populous districts of Broughton, Higher and Lower, did not exist thirty-five

* The corner of Irwell Street, Salford, as recently as 1828, was occupied by an old canal "flat," tenanted by an eccentric character, after whom it was designated "Bannister's Ship." Opposite it was a row of cottages with gardens in front. Oldfield and Ordsall Lanes were country roads, and the streets adjacent to them were not yet in existence.

years ago. They contained no buildings excepting Strangeways Hall and a few cottages which lay scattered beyond the bottom of the workhouse brow; the locality where the new Assize Courts have been erected, which the citizens of Manchester claim to be unequalled in the kingdom for magnificence and accommodation.

But pastures, corn-fields, and gardens rapidly gave place to streets and factory buildings.* The suburban districts of Ardwick, Hulme, and Cheetham, became wholly absorbed in the great city. Stretford New Road, a broad street nearly a mile and a half long, forms the main highway for a district occupied during the life of the present generation by a population greater than that of many cities. Not fifty years since, a few farm-houses and detached dwellings were all the buildings it contained, and Chester Road, the principal one in the district, was a narrow winding lane, with hedges on each side. Jackson's Lane, remembered as a mere farm-road through corn-fields, has become a spacious thoroughfare dignified with the name of Great Jackson Street, that contains a relic of rural Hulme in the remnant of " Jackson's Farm" buildings, which gave the name, first to the "lane" and then to the "street." It is a single-storey building, covered with grey flags, and stands in an oblique recess on the left-hand side, about half-way between Chester Road and the recently formed City Road. Higher up, at the junction of Chapman

---

* The growth of Manchester, and the sister borough of Salford, will be more readily appreciated, perhaps, by a glance at the population at different periods than by any other illustration :—

Population in	1774.	1801.	1821.	1861.
	41,032	84,020	187,031	460,028

and Preston Streets, the houses, also covered with grey flags, still remain, which, within a comparatively recent period, stood amidst fields, and were known as "Geary's Farm,"—these buildings are now surrounded by streets on every side. About thirty years since, the part of Hulme nearest to Manchester was occupied by "tea gardens," and places of resort much used by the "town" population. The principal of these was at the White House; and it is said of late roysterers at that place, that unless they could form a party or secure the services of "the patrol," they had frequently to sojourn there all night. The officers constituting the patrol * carried swords and horn lanterns; and, clad as they were in heavy greatcoats with many capes, they were by no means light of foot, or at all formidable adversaries to the footpads who "worked" the district.

Among the most remarkable improvements in Manchester of late years, have been the numerous spacious thoroughfares which have been opened up in all directions. In this respect, the public spirit of Manchester has not been surpassed by any town in the kingdom,—the new streets being laid out on a settled plan with a view to future extension, and executed with admirable judgment. Narrow, dark, and crooked ways have been converted into wide and straight streets, admitting light, air, and health to the inhabitants, and affording spacious highways for the great and growing traffic of the district.

---

* On the subject of watchmen, it may be mentioned that the first watchman was appointed for Chorlton-on-Medlock in 1814. In 1832 an Act was obtained for improving and regulating that township, and so recently as 1833 it was first lighted with gas. The Police Act for the Township of Hulme was obtained in 1834.

The important street-improvements executed in Manchester during the last thirty years have cost an aggregate of about 800,000*l.* The central and oldest part of the town has thus undergone a complete transformation. So numerous are the dark and narrow entries that have been opened up—the obstructive buildings that have been swept away, the projecting angles that have been cut off, and the crooked ways that have been made straight—that the denizen of a former age would be very unlikely to recognise the Manchester of to-day, were it possible for him to revisit it.

Some of the street-improvements have their peculiar social aspects, and call up curious reminiscences. The stocks, pillory, and Old Market Cross, were removed from the Market Place in 1816. The public whipping of culprits on the pillory stage is within the recollection of the elder portion of the present inhabitants. Another "social institution," of a somewhat different character, was extinguished much more recently, by the construction of the splendid piece of terrace-road in front of the cathedral, known as the Hunt's Bank improvement. This road swept away a number of buildings, shown on the old plans of Manchester as standing on the water's edge, close to the confluence of the Irk with the Irwell. They were reached by a flight of some thirty steps, and consisted of a dye-work, employing three or four hands, two public-houses, and about a dozen cottages and other buildings. The public-houses, the 'Ring o' Bells' and the 'Blackamoor,' particularly the former, were famous places in their day. On Mondays, wedding-parties from the country, consisting sometimes of from twenty to thirty couples, accompanied by fiddlers,

visited " t' Owd Church " to get married. The ' Ring
o' Bells ' was the rendezvous until the parties were
duly married and ready to form and depart home-
wards, in a more or less orderly manner, headed by
their fiddlers as they had come.   The ' Ring o' Bells'
was also a favourite resort of the recruiting-serjeant,
and more recruits, it is said, were enlisted there than
at any other public-house in the kingdom.   But
these, and many curious characteristics of old Manchester, have long since passed away ; and not only
the town but its population have become entirely
new.

Bridgewater Halfpenny Token.

CHAPTER VI.

BRINDLEY CONSTRUCTS THE GRAND TRUNK CANAL.

Long before the Duke's Canal was finished, Brindley was actively employed in carrying out a still larger enterprise,—a canal to connect the Mersey with the Trent, and both with the Severn,—thus uniting by a grand line of water-communication the ports of Liverpool, Hull, and Bristol. He had, as we have already seen, made a survey of such a canal, at the instance of Earl Gower, before his engagement as engineer for the Bridgewater undertaking. Thus, in the beginning of February, 1758, before the Duke's bill had been even applied for, we find him occupied for days together "a bout the novogation," and he then surveyed the country between Longbridge in Staffordshire, and King's Mills in Derbyshire.

The enterprise, however, made very little progress. The success of canals in England was as yet entirely problematical ; and this was of too formidable a character to be hastily undertaken. But again, in 1759, we find Brindley proceeding with his survey of the Staffordshire Canal ; and in the middle of the following year he was occupied about twenty days in levelling from Harecastle, at the summit of the proposed canal, to Wilden, near Derby. During that time he had many interviews with Earl Gower at Trentham, and with the Earl of Stamford at Enville, discussing the project.

The next step taken was the holding of a public meeting at Sandon, in Staffordshire, as to the proper course which the canal should take, if finally decided upon. Considerable difference of opinion was expressed at the meeting, in consequence of which it was arranged that Mr. Smeaton should be called upon to co-operate with Brindley in making a joint survey and a joint report. A second meeting was held at Wolseley Bridge, at which the plans of the two engineers were ordered to be engraved and circulated amongst the landowners and others interested in the project. Here the matter rested for several years more, without any further action being taken. Brindley was hard at work upon the Duke's Canal, and the Staffordshire projectors were disposed to wait the issue of that experiment; but no sooner had it been opened, and its extraordinary success become matter of fact, than the project of the canal through Staffordshire was again revived. The gentlemen of Cheshire and Staffordshire, especially the salt manufacturers of the former county and the earthenware-manufacturers of the latter, now determined to enter into co-operation with the leading landowners in concerting the necessary measures with the object of opening up a line of water-communication with the Mersey and the Trent.

The earthenware manufacture, though in its infancy, had already made considerable progress; but, like every other branch of industry in England at that time, its further development was greatly hampered by the wretched state of the roads. Throughout Staffordshire they were as yet, for the

most part, narrow, deep, circuitous, miry, and
inconvenient; barely passable with rude waggons
in summer, and almost impassable, even with pack-
horses, in winter. Yet the principal materials used
in the manufacture of pottery, especially of the best
kinds, were necessarily brought from a great distance—flint-stones from the south-eastern ports of
England, and clay from Devonshire and Cornwall.
The flints were brought by sea to Hull, and the
clay to Liverpool. From Hull the materials were
brought up the Trent in boats to Willington; and
the clay was in like manner brought from Liverpool
up the Weaver to Winsford, in Cheshire. Considerable quantities of clay were also conveyed in
boats from Bristol, up the Severn, to Bridgenorth
and Bewdley. From these various points the
materials were conveyed by land-carriage, mostly
on the backs of horses, to the towns in the Potteries,
where they were worked up into earthenware and
china.

The manufactured articles were returned for
export in the same rude way. Large crates of
pot-ware were slung across horses' backs, and thus
conveyed to their respective ports, not only at
great risk of breakage and pilferage, but also at a
heavy cost. The expense of carriage was not less
than a shilling a ton per mile, and the lowest charge
was eight shillings the ton for ten miles. Besides,
the navigation of the rivers above mentioned was
most uncertain, arising from floods in winter and
droughts in summer. The effect was, to prevent
the expansion of the earthenware manufacture, and
very greatly to restrict the distribution of the lower-
priced articles in common use.

The same difficulty and cost of transport checked

the growth of nearly all other branches of industry, and made living both dear and uncomfortable. The indispensable article of salt, manufactured at the Cheshire Wiches, was in like manner carried on horses' backs all over the country, and reached almost a fabulous price by the time it was sold two or three counties off. About a hundred and fifty pack-horses, in gangs, were occupied in going weekly from Manchester, through Stafford, to Bewdley and Bridgenorth, loaded with woollen and cotton cloth for exportation;* but the cost of the carriage by this mode so enhanced the price, that it is clear that in the case of many articles it must have acted as a prohibition, and greatly checked both production and consumption. Even corn, coal, lime, and iron-stone were conveyed in the same way, and the operations of agriculture, as

* In a curious book published in 1766, by Richard Whitworth, of Balcham Grange, Staffordshire, afterwards Sir Richard Whitworth, member for Stafford, entitled 'The Advantages of Inland Navigation,' he points to the various kinds of traffic that might be expected to come upon the canal then proposed by him, and amongst other items he enumerates the following:—" There are three pot-waggons go from Newcastle and Burslem weekly, through Eccleshall and Newport to Bridgenorth, and carry about eight tons of pot-ware every week, at 3l. per ton. The same waggons load back with ten tons of close goods, consisting of white clay, grocery, and iron, at the same price, delivered on their road to Newcastle. Large quantities of pot-ware are conveyed on horses' backs from Burslem and Newcastle to Bridgenorth and Bewdley for exportation—about one hundred tons yearly, at 2l. 10s. per ton. Two broad-wheel waggons (exclusive of 150 pack-horses) go from Manchester through Stafford weekly, and may be computed to carry 312 tons of cloth and Manchester wares in the year, at 3l. 10s. per ton. The great salt-trade that is carried on at Northwich may be computed to send 600 tons yearly along this canal, together with Nantwich 400, chiefly carried now on horses' backs, at 10s. per ton on a medium."

of manufacture, were alike injuriously impeded. There were no shops then in the Potteries, the people being supplied with wares and drapery by packmen and hucksters, or from Newcastle-under-Lyne, which was the only town in the neighbourhood worthy of the name.

The people of the district in question were quite as rough as their roads. Their manners were coarse, and their amusements brutal. Bull-baiting, cock-throwing, and goose-riding were their favourite sports. When Wesley first visited Burslem, in 1760, the potters assembled to jeer and laugh at him. They then proceeded to pelt him. "One of them," he says, "threw a clod of earth which struck me on the side of the head; but it neither disturbed me nor the congregation." At that time the whole population of the Potteries did not amount to more than about 7000 souls. The villages in which they lived were poor and mean, scattered up and down, and the houses were mostly covered with thatch. Hence the Rev. Mr. Middleton, incumbent of Stone—a man of great shrewdness and quaintness, distinguished for his love of harmless mirth and sarcastic humour—when enforcing the duty of humility upon his leading parishioners, took the opportunity, on one occasion, after the period of which we speak, of reminding them of the indigence and obscurity from which they had risen to opulence and respectability. He said they might be compared to so many sparrows, for that all of them had been hatched *under the thatch.* When the congregation of this gentleman, growing rich, bought an organ and placed it in the church, he persisted in calling it the hurdy-gurdy, and often took occasion to lament the loss of his old psalm-singers.

The people towards the north were no better, nor were those further south. When Wesley preached at Congleton, four years later, he said, " even the poor potters [though they had pelted him] are a more civilised people than the better sort (so called) at Congleton." Arthur Young visited the neighbourhood of Newcastle-under-Lyne in 1770, and found poor-rates high, wages low, and employment scarce. " Idleness," said he, " is the chief employment of the women and children. All drink tea, and fly to the parishes for relief at the very time that even a woman for washing is not to be had. By many accounts I received of the poor in this neighbourhood, I apprehend the rates are burthened for the spreading of laziness, drunkenness, tea-drinking, and debauchery, —the general effect of them, indeed, all over the kingdom." *

Hutton's account of the population inhabiting the southern portion of the same county is even more dismal. Between Hales Owen and Stourbridge was a district usually called the Lie Waste, and sometimes the Mud City. Houses stood about in every direction, composed of clay scooped out into a tenement, hardened by the sun, and often destroyed by the frost. The males were half-naked, the children dirty and hung over with rags. " One might as well look for the moon in a coal-pit," says Hutton, " as for stays or white linen in the City of Mud. The principal tool in business is the hammer, and the beast of burden the ass." †

The district, however, was not without its

* Young's ' Six Months' Tour.' Ed. 1770. Vol. iii., p. 317.
† ' History of Birmingham.' Ed. 1836, p. 24.

sprinkling of public-spirited men who were actively
engaged in devising new sources of employment
for the population ; and, as one of the most effective means of accomplishing this object, opening
up the communications, by road and canal, with
near as well as distant parts of the country. One
of the most zealous of such workers was the illustrious Josiah Wedgwood. He was one of those
indefatigable men who from time to time spring
from the ranks of the common people, and by their
energy, skill, and enterprise, not only practically
educate the working population in habits of industry, but, by the example of diligence and perseverance which they set before them, largely
influence public action in all directions, and contribute in a great measure to form the national
character.

Josiah Wedgwood was born in a humble position
in life ; and though he rose to eminence as a man
of science as well as a manufacturer, he possessed
no greater advantages at starting than Brindley
himself did. His grandfather and granduncle were
both potters, as was also his father Thomas, who
died when Josiah was a mere boy, the youngest of
a family of thirteen children. He began his industrial life as a thrower in a small pot-work, conducted by his elder brother ; and he might have
continued working at the wheel but for an attack
of virulent small-pox, which, being neglected,
led to a disease in his right leg, which in a
great measure unfitted him for following even that
humble employment. When he returned to his
work, most probably before he was sufficiently
recovered from his illness, the pain in his limb was
such that he had to keep it almost constantly rested

upon a stool before him.* The disease continued increasing as he advanced in years, and it was greatly aggravated by an unfortunate bruise, which kept him to his bed for months, and reduced him to the last extremity of debility. At length the disorder reached the knee, and threatened to endanger his life, when amputation was found necessary. During the enforced leisure of his many illnesses arising from this cause, Wedgwood took to reading and thinking, and turned over in his mind the various ways of making a living by his trade, now that he could no longer work at the potter's wheel. It has been no less elegantly than truthfully observed by Mr. Gladstone, that "it is not often that we have such palpable occasion to record our obligations to the small-pox. But in the wonderful ways of Providence, that disease, which came to him as a twofold scourge, was probably the occasion of his subsequent excellence. It prevented him from growing up to be the active, vigorous English workman, possessed of all his limbs, and knowing right well the use of them; but it put him upon considering whether, as he could not be that, he might not be something else, and something greater. It sent his mind inwards; it drove him to meditate upon the laws and secrets of his art. The result was, that he arrived at a perception and a grasp of them which might, perhaps, have been envied, certainly have been owned, by an Athenian potter."†

* The Papers of Mr. Llewellynn Jewitt, F.S.A., on 'Wedgwood and Etruria,' published in the *Art Journal*, 1864, contain many interesting particulars relating to the life and labours of Wedgwood, and are well worthy of perusal.

† 'Wedgwood;' an Address delivered at Burslem, Oct. 26, 1863. By the Right Hon. W. E. Gladstone, Chancellor of the Exchequer. London : Murray.

Wedgwood began operations on his own account by making various ornamental objects out of potter's clay, such as knife-hafts, boxes, and sundry curious little articles for domestic use. He joined

Ivy House, Burslem, Wedgwood's first Pottery.*

in several successive partnerships with other workmen, but made comparatively small progress until he began business for himself, in 1759, in a humble cottage near the Market House in Burslem, known by the name of the Ivy House. He there pursued his manufacture of knife-handles and other small wares, striving at the same time to acquire such a knowledge of practical chemistry as might enable him to improve the quality of his work in respect

* The Ivy House, in which Wedgwood began business on his own account, is the cottage shown on the right-hand of the engraving. The other house is the old 'Turk's Head.'

of colour, glaze, and durability. Success attended
Wedgwood's diligent and persistent efforts, and
he proceeded from one stage of improvement to
another, until at length, after a course of about
thirty years' labour, he firmly established a new
branch of industry, which not only added greatly
to the conveniences of domestic life, but proved
a source of remunerative employment to many
thousand families throughout England.

His trade having begun to expand, an extensive
demand for his articles sprang up, not only in
London, but in foreign countries.* But there was
this great difficulty in his way,—that the roads in
his neighbourhood were so bad that he was at the
same time prevented from obtaining a sufficient
supply of the best kinds of clay and also from disposing of his wares in distant markets. This great
evil weighed heavily upon the whole industry of
the district, and Wedgwood accordingly appears
to have bestirred himself at an early period in his
career to improve the local communications. In
conjunction with several of the leading potters he
promoted an application to Parliament for powers
to repair and widen the road from the 'Red Bull,' at

* Faujas Saint Fond, in his 'Travels in England,' thus writes
respecting Wedgwood's ware:—"Its excellent workmanship, its
solidity, the advantage which it possesses of standing the action of
fire, its fine glaze, impenetrable to acids, the beauty, convenience,
and variety of its forms, and its moderate price, have created a
commerce so active, and so universal, that in travelling from Paris
to St. Petersburg, from Amsterdam to the farthest point of Sweden,
from Dunkirk to the southern extremity of France, one is served
at every inn from English earthenware. The same fine article
adorns the tables of Spain, Portugal, and Italy ; and it provides
the cargoes of ships to the East Indies, the West Indies, and
America."

Lawton, in Cheshire, to Cliff Bank, in Staffordshire. This line, if formed, would run right through the centre of the Potteries, open them to traffic, and fall at either end into a turnpike road.

The measure was, however, violently opposed by the people of Newcastle-under-Lyne, on the ground that the proposed new road would enable waggons and pack-horses to travel north and south from the Potteries without passing through their town. The Newcastle innkeepers acted as if they had a vested interest in the bad roads ; but the bill passed, and the new line was made, stopping short at Burslem. This was, no doubt, a great advantage, but it was not enough. The heavy carriage of clay, coal, and earthenware needed some more convenient means of transport than waggons and roads ; and, when the subject of water communication came to be discussed, Josiah Wedgwood at once saw that a canal was the very thing for the Potteries. Hence he immediately entered with great spirit into the movement again set on foot for the construction of Brindley's Grand Trunk Canal.

The field was not, however, so clear now as it had been before. The success of the Duke's canal led to the projection of a host of competing schemes in the county of Chester, and it appeared that Brindley's Grand Trunk project would have to run the gauntlet of a powerful local opposition. There were two other schemes besides his, which formed the subject of much pamphleteering and controversy at the time, one entering the district by the river Weaver, and another by the Dee. Neither of these proposed to join the Duke of Bridgewater's canal, whereas the Grand Trunk line was laid out so as to run into his at Preston-on-the-Hill near

Runcorn. As the Duke was desirous of placing his navigation—and through it Manchester, Liverpool, and the intervening districts—in connection with the Cheshire Wiches and the Staffordshire Potteries, he at once threw the whole weight of his

Josiah Wedgwood.

support upon the side of Brindley's Grand Trunk. Indeed, he had himself been partly at the expense of its preliminary survey, as we find from an entry in Brindley's memorandum-book, under date the 12th of April, 1762, as follows: "Worsley—Rec^d

from Mr Tho Gilbert for ye Staffordshire survey, on account, 33*l.* 16*s.* 11*d.*"

The Cheshire gentlemen protested against the Grand Trunk scheme, as calculated to place a monopoly of the Staffordshire and Cheshire traffic in the hands of the Duke; but they concealed the fact, that the adoption of their respective measures would have established a similar monopoly in the hands of the Weaver Canal Company, whose line of navigation, so far as it went, was tedious, irregular, and expensive.   Both parties mustered their forces for a Parliamentary struggle, and Brindley exerted himself at Manchester and Liverpool in obtaining support and evidence on behalf of his plan. The following letter from him to Gilbert, then at Worsley, relates to the rival schemes.

" 21 Decr. 1765

" On Tusdey Sr Georg [Warren] sent Nuton in to Manchester to make what intrest he could for Sir Georg and to gather ye old Navogtors togather to meet Sir Georg at Stoperd to make Head a ganst His Grace

" I sawe Docter Seswige who sese Hee wants to see you about pamant of His Land in Cheshire

" On Wednesday ther was not much transpired but was so dark I could carse do aneything

" On Thursdey Wadgwood of Burslam came to Dunham & sant for mee and wee dined with Lord Gree [Grey] & Sir Hare Mainwering and others   Sir Hare cud not ceep His Tamer [temper]   Mr. Wedgwood came to seliset Lord Gree in faver of the Staffordshire Canal & stade at Mrs Latoune all night & I whith him & on frydey sat out to wate on Mr Edgerton to seliset Him   Hee sase Sparrow and others are indavering to gat ye Land owners consants from Hare Castle to Agden

" I have ordered Simcock to ye Langth falls of Sanke Navegacion.

" Ryle wants to have coals sant faster to Alteringham that Hee may have an opertunety dray of ye sale Moor Canal in a bout a weeks time.

" I in tend being back on Tusdy at fardest."

The first public movement was made by the supporters of Brindley's scheme. They held an open meeting at Wolseley Bridge, Staffordshire, on the 30th of December, 1765, at which the subject was fully discussed. Earl Gower, the lord-lieutenant of the county, occupied the chair; and Lord Grey and Mr. Bagot, members for the county, —Mr. Anson, member for Lichfield,—Mr. Thomas Gilbert, the agent for Earl Gower, then member for Newcastle-under-Lyne,—Mr. Wedgwood, and many other influential gentlemen, were present to take part in the proceedings. Mr. Brindley was called upon to explain his plans, which he did to the satisfaction of the meeting; and these having been adopted, with a few immaterial alterations, it was determined that steps should be taken to apply for a bill conferring the necessary powers in the next session of Parliament. Mr. Wedgwood put his name down for a thousand pounds towards the preliminary expenses, and promised to subscribe largely for shares besides.* The promoters of the measure proposed to designate the undertaking " The Canal from the Trent to the Meresy;" but Brindley, with sagacious foresight, urged that it

---

* Wedgwood even entered the lists as a pamphleteer in aid of the Grand Trunk project, and, in 1765, he and his partner, Mr. Bentley, formerly of Liverpool, drew up a very able statement, showing the advantages likely to be derived from the construction of the proposed canal, under the title of ' A View of the Advantages of Inland Navigation, with a plan of a Navigable Canal intended for a communication between the ports of Liverpool and Hull.' It

should be called The Grand Trunk, because, in his judgment, numerous other canals would branch out from it at various points of its course, in like manner as the arteries of the human system branch out from the aorta; and before many years had passed, his anticipations in this respect were fully realised. The Staffordshire potters were greatly pleased with the decision of the meeting, and on the following evening they assembled round a large bonfire at Burslem, and drank the healths of Lord Gower, Mr. Gilbert, and the other promoters of the scheme, with fervent demonstrations of joy.

The opponents of the measure also held meetings, at which they strongly declaimed against the Duke's proposed monopoly, and set forth the superior merits of their respective schemes. One of these was a canal from the river Weaver, by Nantwich, Eccleshall, and Stafford, to the Trent at Wilden Ferry, without touching the Potteries at all. Another was for a canal from the Weaver at Northwich, passing by Macclesfield and Stockport, round to Manchester, thus completely surrounding the Duke's navigation, and preventing its extension southward into Staffordshire or any other part of the Midland districts.

But there was also a strong party opposed to all canals whatever—the party of croakers, who are always found in opposition to improved communications, whether in the shape of turnpike

pointed out in glowing language the advantages to be derived from opening up the internal communications of a country by means of roads, canals, &c.; and showed how the comfort and even the necessity of all classes must be so much better provided for by a reduction in the cost of carriage of useful and necessary commodities.

roads, canals, or railways. They prophesied that if the proposed canals were made, the country would be ruined, the breed of English horses would be destroyed, the innkeepers would be made bankrupts, and the pack-horses and their drivers would be deprived of their subsistence. It was even said that the canals, by putting a stop to the coasting trade, would destroy the race of seamen. It is a fortunate thing for England that it has contrived to survive these repeated prophecies of ruin. But the manner in which our countrymen contrive to grumble their way along the high road of enterprise, thriving and grumbling, is one of the peculiar features in our character which perhaps only Englishmen can understand and appreciate.

It is a curious illustration of the timidity with which the projectors of those days entered upon canal enterprise, that one of their most able advocates, in order to mitigate the opposition of the pack-horse and waggon interest, proposed that "no main trunk of a canal should be carried nearer than within four miles of any great manufacturing and trading town; which distance from the canal would be sufficient to maintain the same number of carriers and to employ almost the same number of horses as before."* But as none of the towns in the Potteries were as yet large manufacturing or trading places, this objection did not apply to them, nor prevent the canals from being carried quite through the centre of what has since become a continuous district of populous manufacturing towns and villages. The vested interests of some of the larger towns were, however, for this reason,

* 'The Advantages of Inland Navigation,' by R. Whitworth. 1766.

preserved, greatly to their own ultimate injury ; and when the canal, to conciliate the local opposition, was so laid out as to leave them at a distance, not many years elapsed before they became clamorous for branches to join the main trunk—but not until the mischief had been done, and a blow dealt to their own trade, in consequence of their being left so far outside the main line of water communication, from which many of them never after recovered.

It is not necessary to describe the Parliamentary contest upon the Grand Trunk Canal Bill. There was the usual muster of hostile interests,—the river navigation companies uniting to oppose the new and rival company—the array of witnesses on both sides, — Brindley, Wedgwood, Gilbert, and many more, giving their evidence in support of their own scheme, and a powerful array of the Cheshire gentry and Weaver Navigation Trustees appearing on behalf of the others, — and the whipping-up of votes, in which the Duke of Bridgewater and Earl Gower worked their influence with the Whig party to good purpose.

Brindley's plan was, on the whole, considered the best. It was the longest and the most circuitous, but it appeared calculated to afford the largest amount of accommodation to the public. It would pass through important districts, urgently in need of an improved communication with the port of Liverpool on the one hand, and with Hull on the other. But it was not so much the connection of those ports with each other that was needed, as a more convenient means of communication between them and the Staffordshire manufacturing districts ; and the Grand Trunk system—somewhat in the form of a horse-shoe, with the Potteries lying

along its extreme convex part—promised effectually
to answer this purpose, and to open up a ready
means of access to the coast on both sides of the
island.

A glance at the course of the proposed line will
show its great importance. Starting from the
Duke's canal at Preston-on-the-Hill, near Runcorn,
it passed southwards by Northwich and Middlewich, through the great salt-manufacturing districts
of Cheshire, to the summit at Harecastle. It was
alleged that the difficulties presented by the long
tunnel at that point were so great that it could
never be the intention of the projectors of the canal
to carry their "chimerical idea," as it was called,
into effect. Brindley however insisted, not only
that the tunnel was practicable, but that, if the
necessary powers were granted, he would certainly execute it.* Descending from the summit
level into the valley of the Trent, the canal proceeded southwards through the Pottery districts,
passing close to Burslem, Hanley, Stoke, and
Lane End. It then passed onward, still south, by

---

* In one of the many angry pamphlets published at the time,
the 'Supplement to a pamphlet entitled Seasonable Considerations
on a Navigable Canal intended to be cut from the Trent to the
Mersey,' &c., the following passage occurs : "When our all is at
stake, these gentlemen [the promoters of the Grand Trunk Canal]
must not be surprised at bold truths. We conceive more favourably of their *understanding* than of their *motive;* we cannot
suspect them of entertaining the chimerical idea of cutting through
Hare Castle ! We rather believe that they are desirous of cutting
their canal at both ends, and of leaving the middle for the project
of a future day. Are these projectors *jealous* of their *honour ?*
Let them adopt a clause (which reason and justice strongly enforce)
to restrain them from meddling with *either end* till they have
finished the *great trunk.* This, and this alone, will shield them
from suspicion."

Trentham, Stone, and Shutborough, to Haywood, where it joined the canal projected to unite the Severn with the Mersey. Still following the valley of the Trent, the canal near Rugeley, turning sharp round, proceeded in a north-easterly direction, nearly parallel with the river, passing Burton and Ashton, to a junction with the main stream at Wilden Ferry, a little above where the Derwent falls into the Trent near Derby.

From thence there was a clear line of navigation, by Nottingham, Newark, and Gainsborough, to the Humber.

Provided this admirable project could be carried out, it appeared likely to meet all the necessities of the case. Ample evidence was given in support of the allegations of its promoters; and the result was, that Parliament threw out the bills promoted by the Cheshire gentlemen on behalf of the old river navigation interest, and the Grand Trunk Canal Act passed into law. At the same time another important Act was passed, empowering the construction of the Wolverhampton Canal, from the river Severn, near Bewdley, to the river Trent, near Haywood Mill; thus uniting the navigation of the three rivers which had their termini at the ports of Liverpool, Hull, and Bristol, on the opposite sides of the island.

There was great rejoicing at Burslem on the news arriving at that place of the passing of the bill; and very shortly after, on the 26th of July, 1766, the works were formally commenced by Josiah Wedgwood on the declivity of Bramhills, in a piece of land within a few yards of the bridge which crosses the canal at that place. Brindley was present at the ceremony, when due honours were paid him by the assembled potters. After Mr. Wedgwood had cut the first sod, many of the leading persons of the neighbourhood followed his example, putting their hand to the work by turns, and each cutting a turf or wheeling a barrow of earth in honour of the occasion. It was, indeed, a great day for the Potteries, as the event proved. In the afternoon a sheep was roasted whole in Burslem market-place, for the good of the poorer class of potters; a *feu de joie* was fired in front of Mr. Wedgwood's house, and sundry other demonstrations of local rejoicing wound up the day's proceedings.

Wedgwood was of all others the most strongly impressed with the advantages of the proposed canal. He knew and felt how much his trade had been hindered by the defective communications of the neighbourhood, and to what extent it might be increased provided a ready means of transit to Liverpool, Hull, and Bristol could be secured; and, confident in the accuracy of his anticipations, he procceded to make the purchase of a considerable estate in Shelton, intersected by the canal, on the banks of which he built the celebrated Etruria—the finest manufactory of the kind up to that time erected in England, alongside of which he built a mansion for himself and cottages for his work-people. He removed his works thither from Burslem, partially in 1769, and wholly in 1771, shortly before the works of the canal had been completed.

The Grand Trunk was the most formidable undertaking of the kind that had yet been attempted in England. Its whole length, including the junctions with the Birmingham Canal and the river Severn, was 139½ miles. In conformity with Brindley's practice, he laid out as much of the navigation as possible upon a level, concentrating the locks in this case at the summit, near Hare-castle, from which point the waters fell in both directions, north and south. Brindley's liking for long flat reaches of dead water made him keep clear of rivers as much as possible. He likened water in a river flowing down a declivity, to a furious giant running along and overturning everything; whereas (said he) "if you lay the giant flat upon his back, he loses all his force, and becomes completely passive, whatever his size may be." Hence he contrived that from Middlewich, a distance of seventeen miles, to the Duke's Canal at Preston Brook, there should not be a lock; but goods might be conveyed from the centre of Cheshire to Manchester, for a distance of about seventy miles, along one uniform water level. He carried out the same practice, in like manner, on the Trent side of Harecastle, where he laid out the canal in as many long lengths of dead water as possible.

The whole rise of the canal from the level of the Mersey, including the Duke's locks at Runcorn, to the summit at Harecastle, is 395 feet; and the fall from thence to the Trent at Wilden is 288 feet 8 inches. The locks of the Grand Trunk proper, on the northern side of Harecastle, are thirty-five, and on the southern side forty. The dimensions of the canal, as originally constructed, were twenty-eight feet in breadth at the top, sixteen at the bottom, and four and a half feet in depth; but from Wilden to Burton, and from Middlewich to Preston-on-the-Hill, it was thirty-one feet broad at the top, eighteen at the bottom, and five and a half feet deep, so as to be navigable by large barges; and the locks at those parts of the canal were of correspondingly large dimensions. The width was afterwards made uniform throughout. The canal was carried over the river Dove on an aqueduct of twenty-three arches, approached by an embankment on either side—in all a mile and two furlongs in length. There were also aqueducts over the Trent, which it crosses at four different points—one of these being of six arches of twenty-one feet span each—and over the Dane and other smaller streams. The number of minor aqueducts was about 160, and of road-bridges 109.

I.   Y

But the most formidable works on the canal
were the tunnels, of which there were five—the
Harecastle, 2880 yards long ; the Hermitage, 130
yards ; the Barnton, 560 yards ; the Saltenford,
350 yards ; and the Preston-on-the-Hill, 1241 yards.
The Harecastle tunnel (subsequently duplicated by
Telford) was constructed only nine feet wide and
twelve feet high ; * but the others were seventeen
feet four inches high, and thirteen feet six inches
wide. The most extensive ridge of country to be
penetrated was at Harecastle, involving by far
the most difficult work in the whole undertaking.
This ridge is but a continuation of the high
ground, forming what is called the "back-bone of
England," which extends in a south-westerly direction from the Yorkshire mountains to the Wrekin
in Shropshire. The flat county of Cheshire, which
looks almost as level as a bowling-green when
viewed from the high ground near New Chapel,
seems to form a deep bay in the land, its innermost
point being almost immediately under the village
of Harecastle ; and from thence to the valley of the
Trent the ridge is at the narrowest. That Brindley
was correct in determining to form his tunnel at
this point has since been confirmed by the survey
of Telford, who there constructed his parallel
tunnel for the same canal, and still more recently

---

* Brindley's tunnel had only space for a narrow canal-boat to
pass through, and it was propelled by the tedious and laborious
process of what is called "legging." It still continues to be worked
in the same way, while horses haul the boats through the whole
length of Telford's wider tunnel. The men who "leg" the boat,
literally kick it along from one end to the other. They lie on their
backs on the boat-cloths, with their shoulders resting against some
package, and propel it along by means of their feet pressing
against the top or sides of the tunnel.

by the engineers of the North Staffordshire Railway, who have also formed their railway tunnel almost parallel with the line of both canals.

When Brindley proposed to cut a navigable way under this ridge, it was declared to be chimerical in the extreme. The defeated promoters of the rival projects continued to make

Northern Entrance of Harecastle Tunnels.*

war upon it in pamphlets, and in the exasperating language of mock sympathy proclaimed Brindley's proposed tunnel to be "a sad misfortune," †

* The smaller opening into the hill on the right-hand of the view is Brindley's tunnel; that on the left is Telford's, executed some forty years since. Harecastle church and village occupy the ground over the tunnel entrances.

† 'Seasonable Considerations,' &c.; Canal pamphlet dated 1766.

inasmuch as it would utterly waste the capital raised by the subscribers, and end in the inevitable ruin of the concern. Some of the small local wits spoke of it as another of Brindley's "Air Castles;" but the allusion was not a happy one, as his first "castle in the air," despite all prophecies to the contrary, had been built, and continued to stand firm at Barton; and judging by the issue of that undertaking, it was reasonable to infer that he might equally succeed in this, difficult though it was on all hands admitted to be.

The Act was no sooner passed than Brindley set to work to execute the impossible tunnel. Shafts were sunk from the hill-top at different points down to the level of the intended canal. The stuff was drawn out of the shafts in the usual way by horse-gins; and so long as the water was met with in but small quantities, the power of windmills and watermills working pumps over each shaft was sufficient to keep the excavators at work. But as the miners descended and cut through the various strata of the hill on their downward progress, water was met with in vast quantities; and here Brindley's skill in pumping machinery proved of great value. The miners were often drowned out, and as often set to work again by his mechanical skill in raising water. He had a fire-engine, or atmospheric steam-engine, of the best construction possible at that time, erected on the top of the hill, by the action of which great volumes of water were pumped out night and day.

This abundance of water, though it was a serious hindrance to the execution of the work, was a circumstance on which Brindley had calculated, and indeed depended, for the supply of water for the summit level of his canal. When the shafts had been sunk to the proper line of the intended waterway, the excavation then proceeded in opposite directions, to meet the other driftways which were in progress. The work was also carried forward at both ends of the tunnel, and the whole line of excavation was at length united by a continuous driftway—it is true, after long and expensive labour—when the water ran freely out at both ends, and the pumping apparatus on the hilltop was no longer needed. At a general meeting of the Company, held on the 1st October, 1768, after the works had been in progress about two years, it appeared from the report of the Committee that four hundred and nine yards of the tunnel were cut and vaulted, besides the vast excavations at either end for the purpose of reservoirs; and the Committee expressed their opinion that the work would be finished without difficulty.

Active operations had also been in progress at other parts of the canal. About six hundred men in all were employed, and Brindley went from point to point superintending and directing their labours. A Burslem correspondent, in September, 1767, wrote to a distant friend thus:—"Gentlemen come to view our eighth wonder of the world, the subterraneous navigation, which is cutting by the great Mr. Brindley, who handles rocks as easily as you would plum-pies, and makes the four elements subservient to his will. He is as plain a looking man as one of the boors of the Peak, or as one of his own carters; but when he speaks, all ears listen, and every mind is filled with wonder at the things he pronounces to be practicable. He has cut a mile through bogs, which he binds up, embanking them with stones which he gets out of other parts of the navigation, besides about a quarter of a mile into the hill Yelden, on the side of which he has a pump worked by water, and a stove, the fire of which sucks through a pipe the damps that would annoy the men who are cutting towards the centre of the hill. The clay he cuts out serves for bricks to arch the subterraneous part, which we heartily wish to see finished to Wilden Ferry, when we shall be able to send Coals and Pots to London, and to different parts of the globe."

In the course of the first two years' operations, twenty-two miles of the navigation had been cut and finished, and it was expected that before eighteen months more had elapsed the canal would be ready for traffic by water between the Potteries and Hull on the one hand, and Bristol on the other. It was also expected that by the same time the canal would be ready for traffic from the north end of Harecastle Tunnel to the river Mersey. The execution of the tunnel, however, proved so tedious and difficult, and the excavation and building went on so slowly, that the Committee could not promise that it would be finished in less than five years from that time. As it was, the completion of the Harecastle Tunnel occupied nine years more; and it was not finished until the year 1777, by which time the great engineer had finally rested from all his labours.

It is scarcely necessary to describe the benefits which the canal conferred upon the inhabitants of the districts through which it passed. As we have already seen, Staffordshire and the adjoining counties had been inaccessible during the chief part of each year. The great natural wealth which they contained was of little value, because it could with difficulty be got at ; and even when reached, there was still greater difficulty in distributing it. Coal could not be worked at a profit, the price of land-carriage so much restricting its use, that it was placed altogether beyond the reach of the great body of consumers.

It is difficult now to realise the condition of poor people situated in remote districts of England less than a century ago. In winter time they shivered over scanty wood-fires, for timber was almost as scarce and as dear as coal. Fuel was burnt only at cooking times, or to cast a glow about the hearth in the winter evenings. The fireplaces were little apartments of themselves, sufficiently capacious to enable the whole family to sit within the wide chimney, where they listened to stories or related to each other the events of the day. Fortunate were the villagers who lived hard by a bog or a moor, from which they could cut peat or turf at will. They ran all risks of ague and fever in summer, for the sake of the ready fuel in winter. But in places remote from bogs, and scantily timbered, existence was scarcely possible ; and hence the settlement and cultivation of the country were in no slight degree retarded until comparatively recent times, when better communications were opened up.

When the canals were made, and enabled coals to be readily conveyed along them at comparatively moderate rates, the results were immediately felt in the increased comfort of the people. Employment became more abundant, and industry sprang up in their neighbourhood in all directions. The Duke's canal, as we have seen, gave the first great impetus to the industry of Manchester and that district. The Grand Trunk had precisely the same effect throughout the Pottery and other districts of Staffordshire ; and their joint action was not only to employ, but actually to civilise the people. The salt of Cheshire could now be manufactured in immense quantities, readily conveyed away, and sold at a comparatively moderate price in all the midland districts of England. The potters of Burslem and Stoke, by the same mode of conveyance, received their gypsum from Northwich, their clay and flints from the seaports now directly connected with the canal, and returned their manufactures by the same route. The carriage of all articles being reduced to about one-fourth of their previous rates,* articles of necessity and comfort, such as had formerly been unknown except amongst the wealthier classes, came into common use amongst the people. Employment increased, and the difficulties of subsistence diminished. Led by the enterprise of Wedgwood and others like

* The following comparison of the rates per ton at which goods were conveyed by land-carriage before the opening of the Grand Trunk Canal, and those at which they were subsequently carried by it, will show how great was the advantage conferred on the country by the introduction of navigable canals :—" The cost of carrying a ton of goods from Liverpool to Etruria, the centre of the Staffordshire Potteries, by land-carriage, was 50s.; the Trent and Mersey reduced it to 13s. 4d. The land-carriage from Liverpool to Wolverhampton was 5l. a ton ; the canal reduced it to 1l. 5s. The land-carriage from Liverpool to Birmingham, and also to Stourport, was 5l. a ton ; the canal reduced both to 1l. 10s. . . . . Thus the cost of inland transport was reduced, on the average, to

him, new branches of industry sprang up, and the
manufacture of earthenware, instead of being insignificant and comparatively unprofitable, which
it was before his time, became a staple branch of
English trade. Only about ten years after the
Grand Trunk Canal had been opened, Wedgwood
stated in evidence before the House of Commons,
that from 15,000 to 20,000 persons were then employed in the earthenware-manufacture alone,
besides the large number of labourers employed
in digging coals for their use, and the still larger
number occupied in providing materials at distant
parts, and in the carrying and distributing trade
by land and sea. The annual import of clay and
flints into Staffordshire at that time was from fifty
to sixty thousand tons ; and yet, as Wedgwood
truly predicted, the trade was but in its infancy.
The tonnage outwards and inwards at the Potteries
is now upwards of three hundred thousand tons
a-year.

The moral and social influences exercised by
the canals upon the Pottery districts were not less
remarkable. From a half-savage, thinly-peopled
district of some 7000 persons in 1760, partially
employed and ill-remunerated, we find them
increased, in the course of some twenty-five
years, to about treble the population, abundantly

about one-fourth of the rate paid previous to the introduction of
canal navigation. The advantages were enormous : wheat, for
example, which formerly could not be conveyed a hundred
miles, from corn-growing districts to the large towns and manufacturing districts, for less than 20s. a quarter, could be conveyed
for about 5s. a quarter. These facts show how great was the
service conferred on the country by Brindley and the Duke of
Bridgewater."—Baines's 'History of the Commerce and Town of
Liverpool.'

employed, prosperous, and comfortable.* Civilisation is doubtless a plant of very slow growth, and
does not necessarily accompany the rapid increase
of wealth.   On the contrary, higher earnings,
without improved morale, may only lead to wild
waste and gross indulgence.   But the testimony
of Wesley to the improved character of the population of the Pottery district in 1781, within a few
years after the opening of Brindley's Grand Trunk
Canal, is so remarkable, that we cannot do better
than quote it here; and the more so, as we have
already given the account of his first visit in 1760,
on the occasion of his being pelted.   "I returned
to Burslem," says Wesley; "how is the whole
face of the country changed in about twenty years !
Since which, inhabitants have continually flowed in
from every side.   Hence the wilderness is literally
become a fruitful field.   Houses, villages, towns,
have sprung up, and the country is not more
improved than the people."

* The population of the same district in 1861 was found to be
upwards of 120,000.

CHAPTER VII.

BRINDLEY'S LAST CANALS—HIS DEATH AND CHARACTER.

IT is related of Brindley that, on one occasion, when giving evidence before a Committee of the House of Commons, in which he urged the superiority of canals over rivers for purposes of inland navigation, the question was asked by a member, " Pray, Mr. Brindley, what then do you think is the use of navigable rivers?" " To make canal navigations, to be sure," was his instant reply. It is easy to understand the gist of the engineer's meaning. For purposes of trade he regarded regularity and certainty of communication as essential conditions of any inland navigation; and he held that neither of these could be relied upon in the case of rivers, which are in winter liable to interruption by floods, and in summer by droughts. In his opinion, a canal, with enough of water always kept banked up, or locked up where the country would not admit of the level being maintained throughout, was absolutely necessary to satisfy the requirements of commerce. Hence he held that one of the great uses of rivers was to furnish a supply of water for canals. It was only another illustration of the "nothing like leather" principle; Brindley's head being so full of canals, and his labours so much confined to the making of canals, that he could think of little else.

In connection with the Grand Trunk—which proved, as Brindley had anticipated, to be the great aorta of the canal system of the midland districts of England—numerous lines were projected and afterwards carried out under our engineer's superintendence. One of the most important of these was the Wolverhampton Canal, connecting the Trent with the Severn, and authorised in the same year as the Grand Trunk itself. It is now known as the Staffordshire and Worcestershire Canal, passing close to the towns of Wolverhampton and Kidderminster, and falling into the Severn at Stourport. This branch opened up several valuable coal-fields, and placed Wolverhampton and the intermediate districts, now teeming with population and full of iron manufactories, in direct connection with the ports of Liverpool, Hull, and Bristol. Two years later, in 1768, three more canals, laid out by Brindley, were authorised to be constructed : the Coventry Canal to Oxford, connecting the Grand Trunk system by Lichfield with London and the navigation of the Thames ; the Birmingham Canal, which brought the advantages of inland navigation to the very doors of the central manufacturing town in England ; and the Droitwich Canal, to connect that town by a short branch with the river Severn. In the following year a further Act was obtained for a canal laid out by Brindley, from Oxford to the Coventry Canal at Longford, eighty-two miles in length.

These were highly important works ; and though they were not all carried out strictly after Brindley's plans, they nevertheless formed the groundwork of future Acts, and laid the foundations of the midland canal system. Thus, the

Coventry Canal was never fully carried out after
Brindley's designs; a difference having arisen
between the engineer and the Company during
the progress of the undertaking, in consequence,
as is supposed, of the capital provided being altogether inadequate to execute the works considered
by Brindley as indispensable. He probably foresaw that there would be nothing but difficulty, and
very likely there might be discredit attached to
himself by continuing connected with an undertaking the proprietors of which would not provide
him with sufficient means for carrying it forward
to completion; and though he finished the first
fourteen miles between Coventry and Atherstone,
he shortly after gave up his connection with the
undertaking, and it remained in an unfinished
state for many years, in consequence of the
financial difficulties in which the Company had
become involved through the insufficiency of their
capital. The connection of the Coventry Canal with
the Grand Trunk was afterwards completed, in 1785,
by the Birmingham and Fazely and Grand Trunk
Companies conjointly, and the property eventually
proved of great value to all parties concerned.

The Droitwich Canal, though only a short branch
five and a half miles in length, was a very important work, opening up as it did an immense trade
in coal and salt between Droitwich and the Severn.
The works of this navigation were wholly executed
by Brindley, and are considered superior to those
of any others on which he was engaged. Whilst
residing at Droitwich, we find our engineer actively
engaged in pushing on the subscription to the Birmingham Canal, the capital of which was taken
slowly. Matthew Boulton, of Birmingham, was one

of the most active promoters of the scheme, and
Josiah Wedgwood also bestirred himself in its behalf.
In a letter written by him about this time, we find
him requesting one of his agents to send out plans
to gentlemen whom he names, in the hope of completing the subscription list.* Brindley did not
live to finish the Birmingham Canal ; it was carried .
out by his successors,—partly by his pupil, Mr.
Whitworth, and partly by Smeaton and Telford.
Brindley's plan was, as usual, to cut the canal as
flat as possible, to avoid the necessity for lockage ;
but his successors, in order to relieve the capital
expenditure, as they supposed, constructed it with
a number of locks to carry it over the summit at
Smithwick. Shortly after its opening, however,
the Company found reason to regret their rejection

---

* The letter is so characteristic of Josiah Wedgwood that we
here insert it at length, as copied from the original in the possession of Mr. Mayer of Liverpool :—

"Burslem, 12th July, 1769.

"Dear Sir,—I should have wrote to you about young Wilson, but the
multiplicity of branches you wrote me he was expected to learn, made me
despair of teaching him any. Pray give my compliments to his father, and if
he chooses to have his son to learn to be a warehouseman and book-keeper,
which is quite sufficient and better than more for any one person, I will learn
him those in the best manner ; but, even then, Mr. Wilson must not expect
him to be set on the top of a ladder without setting his feet upon the lowermost
steps ; and unless he will let the Boy pursue that method, I would not be concerned with him on any account. I will not attempt to teach him any more
trades ; it would injure the Boy, and do me no good. If he has a mind at his
leisure time to amuse himself with drawing I have no objection, and would
encourage him in it, as an innocent amusement, and what may be of use to
him, but would not make this a branch of his business. If the business I propose is too humble for Mr. Wilson's son, I would not by any means have him
accept of it.

"Mr. Brindley desires you'll send 30 plans to each of the undermentioned
Gentlemen by the first Waggons, and let us know when they are sent, as we
shall advertise them in several of the Country papers : Mr. Walker, of Oxford,
Steward to D. of Marlbro'—you may perhaps get them sent from Marlbro'
house ; Mr. Dudley, Attorney-at-Law, Coventry ; Mr. Richardson, Silver-
smith, in Chester ; Mr. Perry, Wolverhampton.

"We shall send you this week end, double salts, creams, pott'y potts, table
plates of all sorts, sallad dishes, covered do., pierced desert plates, &c. We
cannot get Sadler to send us any pierced desert ware, &c. Pay Addison 5 or 6
guineas.

"Yours, &c., J. W."

of Mr. Brindley's advice, and they lowered the summit by cutting a tunnel, as he had originally recommended, thereby incurring an extra expense of about 30,000*l.*

Another of Brindley's canals, authorised in 1769, was that between Chesterfield and the river Trent, at Stockwith, about forty-six miles in length, intended for the transport of coal, lime, and lead from the rich mineral districts of Derbyshire, and the return trade of deals, corn, and groceries to the same districts. It would appear that Mr. Grundy, another engineer, of considerable reputation in his day, was consulted about the project, and that he advised a much more direct route than that pointed out by Brindley, who looked to the accommodation of the existing towns, rather than shortness of route, as the main thing to be provided for. Brindley, in this respect, took very much the same view in laying out his canals as was afterwards taken by George Stephenson—a man whom he resembled in many respects—in laying out railways. He would rather go *round* an obstacle in the shape of an elevated range of country, than go *through* it, especially if in going round and avoiding expense he could accommodate a number of towns and villages. Besides, by avoiding the hills and following the course of the valleys, along which the population usually lies, he avoided expense of construction and secured flatness of canal; just as Stephenson secured flatness of railway gradient. Although the length of canal to be worked was longer, yet the cost of tunnelling and lockage was avoided. The population of the district was also fully accommodated, which could not have been accomplished by the more direct route through unpopulated districts

or under barren hills. The proprietors of the
Chesterfield Canal concurred in Brindley's view,
adopting his plan in preference to Grundy's, and it
was accordingly carried into effect. This navigation was, nevertheless, a work of considerable difficulty, proceeding, as it did, across a very hilly
country, the summit tunnel at Hartshill being 2850
yards in extent. Like many of Brindley's other
works projected about this time, it was finished by
his brother-in-law, Mr. Henshall, and opened for
traffic several years after the great engineer's death.*

The whole of these canals were laid out by
Brindley, though they were not all executed by
him, nor precisely after his plans. No record of
any kind has been preserved of the manner in
which the works were carried out. Brindley himself made few reports, and these merely stated
results, not methods ; yet he had doubtless many
formidable difficulties to encounter, and must have
overcome them by the adoption of those ingenious
expedients, varying according to the circumstances
of each case, in which he was always found so
fertile. He had no treasury of past experience, as
recorded in books, to consult, for he could scarcely
read English ; and certainly he could neither read
French nor Italian, in which languages the only

* The following were the canals laid out and principally executed by Brindley :—

	Miles.	furl.	chains.
The Duke's ⌠ Worsley to Manchester ...	10	2	0
Canals ⌡ Longford Bridge to the Mersey below Runcorn	24	1	7
Grand Trunk Canal Proper, from Wilden Ferry to Preston			
Brook	88	7	9
Wolverhampton Canal	46	4	0
Coventry ,, ...	36	7	8
Birmingham ,, ...	24	2	0
Droitwich ,, ...	5	4	9
Oxford ,, ...	82	7	3
Chesterfield ,, ...	46	0	0

engineering works of any value were then written ; nor had he any store of native experience to draw from, he himself being the first English canal engineer of eminence, and having all his methods and expedients to devise for himself.

It would doubtless have been most interesting could we have had some authentic record of this strong original man's struggles with opposition and difficulty, and the means by which he contrived not only to win persons of high station to support him with their influence but also with their purses, at a time when money was comparatively a much rarer commodity than it is now. "That want of records, journals, and memoranda," says Mr. Hughes, "which is ever to be deplored when we seek to review the progress of engineering works, is particularly felt when we have to look back upon those undertakings which first called for the exercise of engineering skill in many new and untried departments. In Brindley's day, the entire absence of experience derived from former works, the obscure position which the engineer occupied in the scale of society, the imperfect communication between the profession in this country and the engineers and works of other countries, and, lastly, the backward condition of all the mechanical arts and of the physical sciences connected with engineering, may all be ranked in striking contrast with the vast appliances which are placed at the command of modern engineers." *

Moreover, the great canal works upon which Brindley was engaged during the later part of his career, were as yet scarcely appreciated as respects

* 'Memoir,' by Samuel Hughes, C.E. Weale's 'Quarterly Papers,' 1844.

the important influences which they were calculated to exercise upon society at large. The only
persons who seem to have regarded them with
interest were far-sighted men like Josiah Wedgwood, who saw in them the means not only of
promoting the trade of his own county, but of de-
veloping the rich natural resources of the kingdom,
and diffusing amongst the people the elements of
comfort, intelligence, and civilisation. The literary
and scientific classes as yet took little or no interest
in them. The most industrious and observant
literary man of the age, Dr. Johnson, though he
had a word to say upon nearly every subject,
never so much as alluded to them, though all
Brindley's canals were finished in Johnson's lifetime, and he must have observed the works in
progress when passing on his various journeys
through the midland districts. The only reference
which he makes to the projects set on foot for
opening up the country by means of better roads,
was to the effect, that whereas there were before
cheap places and dear places, now all refuges were
destroyed for elegant or genteel poverty.

Before leaving this part of the subject, it is
proper to state that during the latter part of
Brindley's life, whilst canals were being projected
in various directions, he was, on many occasions,
called upon to give his opinion as to the plans
which other engineers had prepared. Among the
most important of the new projects on which he
was thus consulted were, the Leeds and Liverpool
Canal; the improvement of the navigation of the
Thames to Reading; the Calder Navigation; the
Forth and Clyde Canal; the Salisbury and South-
ampton Canal; the Lancaster Canal; and the

Andover Canal. Many of these schemes were of great importance in a national point of view. The Leeds and Liverpool Canal, for instance, brought the whole manufacturing district of Yorkshire along the valley of the Aire into communication with Liverpool and the intermediate districts of Lancashire. The advantages of this navigation to Leeds, Bradford, Keighley, and the neighbouring towns, are felt to this day, and their extraordinary prosperity is doubtless in no small degree attributable to the facilities which the canal has provided for the ready conveyance of raw materials and manufactured produce between those places and the towns and seaports of the west. Brindley surveyed and laid out the whole line of this navigation, 130 miles in length, and he framed the estimate on which the Company proceeded to Parliament for their bill. On the passing of the Act in 1768–9, the Directors appointed him their engineer ; but, being almost overwhelmed with other business at the time, and feeling that he could not give the proper degree of personal attention to carrying out so extensive an undertaking, he was under the necessity of declining the appointment. The works were immediately begun at both ends of the canal, and portions were speedily made use of; but the difficulty and expensiveness of the remaining works greatly delayed their execution, and the canal was not finished until the year 1816. Twenty miles, extending from near Bingley to the neighbourhood of Bradford, were opened on 21st March, 1774. A correspondent of 'Williamson's Liverpool Advertiser' thus describes the opening : "From Bingley and about three miles down, the noblest works of the kind that perhaps are to be found in the universe are exhibited, namely, a five-fold, a three-fold, a two-fold, and a single lock, making together a fall of 120 feet; a large aqueduct-bridge of seven arches over the river Aire, and an aqueduct on a large embankment over Shipley valley. Five boats of burden passed the grand lock, the first of which descended through a fall of sixty-six feet in less than twenty-nine minutes. This much-wished-for event was welcomed with ringing of bells, a band of music, the firing of guns by the neighbouring militia, the shouts of spectators, and all the marks of satisfaction that so important an acquisition merits." On the 21st October of the same year the following paragraph appeared:—" The Liverpool end of the canal was opened from Liverpool to Wigan on Wednesday, the 19th instant, with great festivity and rejoicings. The water had been led into the basin the evening before. At nine A.M. the proprietors sailed up the canal in their barge, preceded by another with music, colours flying, &c., and returned to Liverpool about one. They were saluted with two royal salutes of twenty-one guns each, besides the swivels on board the boats, and welcomed with the repeated shouts of the numerous crowds assembled on the banks, who made a most cheerful and agreeable sight. The gentlemen then adjourned to a tent, on the quay, where a cold colla- tion was set out for themselves and their friends. From thence they went in procession to George's coffee-house, where an elegant dinner was provided. The workmen, 215 in number, walked first, with their tools on their shoulders, and cockades in their hats, and were afterwards plentifully regaled at a dinner provided for them. The bells rang all day, and the greatest joy and order prevailed on the occasion."

Brindley being now the recognised head of his profession, and the great authority on all questions of navigation, he was, in 1770, employed by the Corporation of London to make a survey of the Thames above Battersea, with the object of having it improved for purposes of navigation. As usual, Brindley strongly recommended the construction of a canal in preference to carrying on the navigation by the river, where it was liable to be interrupted by the tides and floods, or by the varying deposits of silt in the shallow places. In his first report to the Common Council, dated the 16th of June, 1770, he pointed out that the cost of hauling the barges was greatly in favour of the canal. For example, he stated that the expense of taking a vessel of 100 or 120 tons from Isleworth to Sunning, and back again to Isleworth, was 80*l.*, and sometimes more; whilst the cost by the canal would only be 16*l.* The saving in time would be still greater, for the double voyage might easily be performed in fifteen hours; whereas by the river the boats were sometimes three weeks in going up, and almost as much in coming down. He estimated that there would be a saving to the public of at least 64*l.* on every voyage, besides the saving of time in performing it. After making a further detailed examination of the district, and maturing his views on the whole subject, he sent in a report, accompanied by a profile of the river about seven feet long, which is still to be seen amongst the records of the Corporation of London. His plan was not, however, carried out; the proposal to construct a canal parallel with the Thames having been abandoned so soon as the Grand Junction Canal was undertaken.

These and numerous other schemes in various
parts of the country—at Stockton, at Leeds, at
Cambridge, at Chester, at Salisbury and South-
ampton, at Lancaster, and in Scotland—fully occupied the attention of Brindley; in addition to
which, there was the personal superintendence
which he must necessarily give to the canals in
active progress, and for the execution of which
he was responsible. In fact, there was scarcely
a design of a canal navigation set on foot throughout the kingdom during the later years of his life,
on which he was not consulted, and the plans of
which he did not entirely make, revise, or improve.

In addition to his canal works, Brindley was
also consulted as to the best means of draining
the low lands in different parts of Lincolnshire,
and the Great Level in the neighbourhood of the
Isle of Ely. He supplied the corporation of
Liverpool with a plan for cleansing the docks
and keeping them clear of mud, which is said to
have proved very effective; and he pointed out
to them an economical method of building walls
against the sea without mortar, which long continued to be employed with complete success. In
such cases he laid his plans freely open to the
public, not seeking to secure them by patent, nor
shrouding his proceedings in any mystery. He was
perfectly open with professional men, harbouring
no petty jealousy of rivals. His pupils had free
access to all his methods, and he took a pride in so
training them that they should reflect credit on the
engineer's profession, then rising into importance,
and be enabled, after he left the scene, to carry on
those great industrial enterprises which he probably
foresaw clearly enough in England's future.

It will be observed, from what we have said, that Brindley's engagements as an engineer extended over a very wide district. Even before his employment by the Duke of Bridgewater, he was under the necessity of travelling great distances to fit up water-mills, pumping-engines, and manufacturing machinery of various kinds, in the counties of Stafford, Cheshire, and Lancashire. But when he had been appointed to superintend the construction of the Duke's canals, his engagements necessarily became of a still more engrossing character, and he had very little leisure left to devote to the affairs of private life. He lived principally at inns, in the immediate neighbourhood of his work; and though his home was at Leek, he sometimes did not visit it for weeks together.

Brindley had very little time for friendship, and still less for courtship. Nevertheless, he did contrive to find time for marrying, though at a comparatively advanced period of his life. In laying out the Grand Trunk Canal, he was necessarily brought into close connection with Mr. John Henshall, of the Bent, near New Chapel, land-surveyor, who assisted him in making the survey. He visited Henshall at his house in September, 1762, and then settled with him the preliminary operations. During his visits Brindley seems to have taken a special liking for Mr. Henshall's daughter Anne, then a girl at school, and when he went to see her father, he was accustomed to take a store of gingerbread for Anne in his pocket. She must have been a comely girl, judging by the portrait of her as a woman, which we have seen.

In due course of time, the liking ripened into an attachment; and shortly after the girl had

left school, at the age of only nineteen, Brindley
proposed to her, and was accepted. By that time
he was close upon his fiftieth year, so that the
union may possibly have been quite as much a
matter of convenience as of love on his part. He
had now left the Duke's service for the purpose
of entering upon the construction of the Grand
Trunk Canal, and with that object resolved to
transfer his home to the immediate neighbourhood of Harecastle, as well as of his colliery at
Golden Hill. Shortly after the marriage, the old
mansion of Turnhurst fell vacant, and Brindley
with his young wife became its occupants. The
marriage took place on the 8th December, 1765,
in the parish church of Wolstanton, Brindley
being described in the register as "of the parish
of Leek, engineer;" but from that time until the
date of his death his home was at Turnhurst.

The house at Turnhurst was a comfortable,
roomy, old-fashioned dwelling, with a garden and
pleasure-ground behind, and a little lake in front.
It was formerly the residence of the Bellot family,
and is said to have been the last mansion in
England in which a family fool was maintained.
Sir Thomas Bellot, the last of the name, was a
keen sportsman, and the panels of several of the
upper rooms contain pictorial records of some of
his exploits in the field. In this way Sir Thomas
seems to have squandered his estate, and it shortly
after became the property of the Alsager family,
from whom Brindley rented it. A little summer-
house, standing at the corner of the outer court-
yard, is still pointed out as Brindley's office, where
he sketched his plans and prepared his calculations. As for his correspondence, it was nearly all

conducted, subsequent to his marriage, by his wife, who, notwithstanding her youth, proved a most clever, useful, and affectionate partner.

Turnhurst was conveniently near to the works then in progress at Harecastle Tunnel, which was within easy walking distance, whilst the colliery at Golden Hill was only a few fields off. From the elevated ground at Golden Hill, the whole range

Brindley's House at Turnhurst

of high ground may be seen under which the tunnel runs—the populous Pottery towns of Tunstall and Burslem filling the valley of the Trent towards the south. At Golden Hill, Brindley carried out an idea which he had doubtless brought with him from Worsley. He and his partners had an underground canal made from the main line of the Harecastle Tunnel into their coal-mine,

about a mile and a half in length; and by that
tunnel the whole of the coal above that level was
afterwards worked out, and conveyed away for
sale in the Pottery and other districts, to the great
profit of the owners and much to the convenience
of the public.

These various avocations involved a great
amount of labour as well as anxiety, and probably
considerable tear and wear of the vital powers.
But we doubt whether mere hard work ever killed
any man, or whether Brindley's labours, extraordinary though they were, would have shortened
his life, but for the far more trying condition of
the engineer's vocation—irregular living, exposure
in all weathers, long fasting, and then, perhaps,
heavy feeding when the nervous system was exhausted, together with habitual disregard of the
ordinary conditions of physical health. These are
the main causes of the shortness of life of most
of our eminent engineers, rather than the amount
and duration of their labours. Thus the constitu-
tion becomes strained, and is ever ready to break
down at the weakest place. Some violation of the
natural laws more flagrant than usual, or a sudden
exposure to cold or wet, merely presents the
opportunity for an attack of disease which the
ill-used physical system is found unable to resist.

Such an accidental exposure unhappily proved
fatal to Brindley. While engaged one day in
surveying a branch canal between Leek and
Froghall, he got drenched near Ipstones, and
went about for some time in his wet clothes.
This he had often before done with impunity,
and he might have done so again; but, unfortunately, he was put into a damp bed in the inn at

Ipstones, and this proved too much for his con-
stitution, robust though he naturally was. He
became seriously ill, and was disabled from all
further work. Diabetes shortly developed itself,
and, after an illness of some duration, he expired
at his house at Turnhurst, on the 27th of September, 1772, in the fifty-sixth year of his age,
and was interred in the burying-ground at New
Chapel, a few fields distant from his dwelling.

James Brindley was probably one of the most
remarkable instances of self-taught genius to be
found in the whole range of biography. The
impulse which he gave to social activity, and the
ameliorative influence which he exercised upon
the condition of his countrymen, seem out of all
proportion to the meagre intellectual culture which
he had received in the course of his laborious and
active career. We must not, however, judge him
merely by the literary test. It is true, he could
scarcely read, and he was thus cut off, to his own
great loss, from familiar intercourse with a large
class of cultivated minds, living and dead; for he
could with difficulty take part in the conversation of educated men, and he was unable to profit
by the rich stores of experience treasured up in
books. Neither could he write, except with difficulty and inaccurately, as we have shown by the
extracts above quoted from his note-books, which
are still extant.
    Brindley was, nevertheless, a highly-instructed
man in many respects. He was full of the results
of careful observation, ready at devising the best
methods of overcoming material difficulties, and
possessed of a powerful and correct judgment in

matters of business. When any emergency arose,
his quick invention and ingenuity, cultivated by
experience, enabled him almost at once unerringly
to suggest the best means of providing for it. His
ability in this way was so remarkable, that those
about him attributed the process by which he
arrived at his conclusions rather to instinct than
reflection—the true instinct of genius. " Mr.
Brindley," said one of his contemporaries, " is one
of those great geniuses whom Nature sometimes
rears by her own force, and brings to maturity
without the necessity of cultivation. His whole
plan is admirable, and so well concerted that he
is never at a loss ; for, if any difficulty arises, he
removes it with a facility which appears so much
like inspiration, that you would think Minerva was
at his fingers' ends."

His mechanical genius was indeed most highly
cultivated. From the time when he bound himself
apprentice to the trade of a millwright—impelled
to do so by the strong bias of his nature—he had
been undergoing a course of daily and hourly
instruction. There was nothing to distract his
attention, or turn him from pursuing his favourite
study of practical mechanics. The training of his
inventive faculty and constructive skill was, indeed,
a slow but a continuous process ; and when the
time and the opportunity arrived for turning these
to account—when the silk-throwing machinery of
the Congleton mill, for instance, had to be perfected and brought to the point of effectively performing its intended work—Brindley was found
able to take it in hand and carry out the plan,
when even its own designer had given it up in
despair. But it must also be remembered that this

extraordinary ability of Brindley was in a great measure the result of close observation, painstaking study of details, and the most indefatigable industry.

The same qualities were displayed in his improvements of the steam-engine, and his arrangements to economise power in the pumping of water from drowned mines. It was often said of his works, as was said of Columbus's discovery, "How easy! how simple!" but this was after the fact. Before he had brought his fund of experience and clearness of vision to bear upon a difficulty, every one was equally ready to exclaim "How difficult! how absolutely impracticable!" This was the case with his "castle in the air," the Barton Viaduct— such a work as had never before been attempted in England, though now any common mason would undertake it. It was Brindley's merit always to be ready with his simple, practical expedient; and he rarely failed to effect his purpose, difficult although at first sight its accomplishment might seem to be.

Like men of a similar stamp, Brindley had great confidence in himself and in his powers and resources. Without this, it had been impossible for him to have accomplished so much as he did. It is said that the King of France, hearing of his wonderful genius, and the works he had performed for the Duke of Bridgewater at Worsley, expressed a desire to see him, and sent a message inviting him to view the Grand Canal of Languedoc. But Brindley's reply was characteristic: "I will have no journeys to foreign countries," said he, "unless to be employed in surpassing all that has been already done in them."

His observation was remarkably quick. In surveying a district, he rapidly noted the character of the country, the direction of the hills and the valleys, and, after a few journeys on horseback, he clearly settled in his mind the best line to be selected for a canal, which almost invariably proved to be the right one. In like manner he would estimate with great rapidity the fall of a brook or river while walking along the banks, and thus determined the height of his cuttings and embankments, which he afterwards settled by a more systematic survey. In these estimates he was rarely, if ever, found mistaken.

His brother-in-law, Mr. Henshall, has said of him, " when any extraordinary difficulty occurred to Mr. Brindley in the execution of his works, having little or no assistance from books or the labours of other men, his resources lay within himself. In order, therefore, to be quiet and uninterrupted whilst he was in search of the necessary expedients, he generally retired to his bed; * and he has been known to be there one, two, or three days, till he had attained the object in view. He would then get up and execute his design, without any drawing or model. Indeed, it was never his custom to make either, unless he was obliged to do it to satisfy his employers. His memory was so remarkable that he has often declared that he could remember, and execute, all the parts of the most complex machine, provided he had time, in his survey of it, to settle in his mind the several parts and their relations to each other. His method of calculating the powers of

---

* The younger Pliny seems to have adopted almost a similar method : " Clausæ fenestræ manent. Mirè enim silentio et tenebris animus alitur. Ab iis quæ avocant abductus, et liber, et mihi relictus, non oculos animo sed animum oculis sequor, qui eadem quæ mens vident quoties non vident alia."—Epist. lib. ix., ep. 36.

any machine invented by him was peculiar to himself. He worked the question for some time in his head, and then put down the results in figures. After this, taking it up again at that stage, he worked it further in his mind for a certain time, and set down the results as before. In the same way he still proceeded, making use of figures only at stated parts of the question. Yet the ultimate result was generally true, though the road he travelled in search of it was unknown to all but himself, and perhaps it would not have been in his power to have shown it to another." *

The statement about his taking to bed to study his more difficult problems is curiously confirmed by Brindley's own note-book, in which he occasionally enters the words "lay in bed," as if to mark the period, though he does not particularise the object of his thoughts on such occasions. It was a great misfortune for Brindley, as it must be to every man, to have his mental operations confined exclusively within the limits of his profession. Anthony Trollope well observes, that "industry is a good thing, and there is no bread so sweet as that which is eaten in the sweat of a man's brow; but the sweat that is ever running makes the bread bitter." Brindley thought and lived mechanics, and never rose above them. He found no pleasure in anything else; amusement of every kind was distasteful to him; and his first

---

* 'Biographia Britannica,' 2nd Ed. Edited by Dr. Kippis. The materials of the article are acknowledged to have been obtained principally from Mr. Henshall by Messrs. Wedgwood and Bentley, who wrote and published the memoir in testimony of their admiration and respect for their deceased friend, the engineer of the Grand Trunk Canal.

visit to the theatre, when in London, was also his last. Shut out from the humanising influence of books, and without any taste for the politer arts, his mind went on painfully grinding in the mill of mechanics. "He never seemed in his element," said his friend Bentley, "if he was not either planning or executing some great work, or conversing with his friends upon subjects of importance." To the last he was full of projects and full of work; and then the wheels of life came to a sudden stop, when he could work no longer.

It is related of him that, when dying, some eager canal undertakers insisted on having an interview with him. They had encountered a serious difficulty in the course of constructing their canal, and they *must* have the advice of Mr. Brindley on the subject. They were introduced to the apartment where he lay scarce able to gasp, yet his mind was clear. They explained their difficulty —they could not make their canal hold water. "Then puddle it," said the engineer. They explained that they had already done so. "Then puddle it again—and again." This was all he could say, and it was enough.

It remains to be added that, in his private character, Brindley commanded general respect and admiration. His integrity was inflexible; his manner, though rough and homely, was kind; and his conduct unimpeachable.* He was altogether

---

* It has, indeed, been stated in the strange publication of the last Earl of Bridgewater, to which we have already alluded, that when in the service of the Duke, Brindley was "drunken." But this is completely contradicted by the testimony of Brindley's own friends; by the evidence of Brindley's note-book, from repeated entries in which it appears that his "ating and drink" at dinner cost no more than 8*d*.; by the confidence generally reposed in him,

unassuming and unostentatious, and dressed and lived with great plainness. His was the furthest possible from a narrow or jealous temper, and nothing gave him greater pleasure than to assist others with their inventions, and to train up a generation of engineers in his pupils, qualified to carry out the works he had himself designed, when he should be no longer able to conduct them. The principal undertakings in which he was engaged up to the time of his death were carried on by his brother-in-law, Mr. Henshall, formerly his clerk of the works on the Grand Trunk Canal, and by his able pupil, Mr. Robert Whitworth, for both of whom he had a peculiar regard, and of whose integrity and abilities he had the highest opinion.

Brindley left behind him two daughters, one of whom, Susannah, married Mr. Bettington, of Bristol, merchant, afterwards the Honourable Mr. Bettington, of Brindley's Plains, Van Diemen's Land, where their descendants still live. His other daughter, Anne, died unmarried, on her passage home from Sydney, in 1838. His widow, still young, married again, and died at Longport in 1826. Brindley had the sagacity to invest a considerable portion of his savings in Grand Trunk shares, the great increase in the value of which, as well as of his colliery property at Golden Hill, enabled him to leave his family in affluent circumstances.

Before finally dismissing the subject of Brindley's

and the friendship entertained for him, by such men as Josiah Wedgwood ; and by the fact of the vast amount of work that he subsequently contrived to get through. No man of "drunken" habits could possibly have done this. We should not have referred to this topic but for the circumstance that the late Mr. Baines, of Leeds, has quoted the Earl's statement, without contradiction, in his excellent ' History of Lancashire.'

I.   2 A canals, we may briefly allude to the influence which they exercised upon the enterprise as well as the speculation of the time. "When these fellows," says Sheridan in the 'Critic,' "have once got hold of a good thing, they do not know when to stop." This might be said of the speculative projectors of canals, as afterwards of railways. The commercial success which followed the opening of the Duke's Canal, and shortly after it the Grand Trunk, soon infected the whole country, and canal schemes were projected in great numbers for the accommodation even of the most remote and unlikely places.

In those districts where the demand for improved water communication grew out of an actual want—as, for instance, where it was necessary to open up a large coal-field for the supply of a population urgently in need of fuel—or where two large towns, such as Manchester and Liverpool, required to be provided with a more cheap and convenient means of trading intercourse than had formerly existed—or where districts carrying on extensive and various manufactures, such as Birmingham, Wolverhampton, and the Potteries, needed a more ready means of communication with other parts of the kingdom—there was no want of trade for the canals; and those constructed for such purposes very soon had as much traffic as they could carry. The owners of land discovered that their breed of horses was not destroyed, and that their estates were not so cut up as to be rendered useless, as many of them had prognosticated. On the contrary, the demand for horses to carry coals, lime, manure, and goods to and from the canal depôts, rapidly increased. The canals meandering through their green fields were no such unsightly objects after all, and they very soon found that inasmuch as the new water-ways readily enabled agricultural produce to reach good markets in the large towns, they were likely even to derive considerable pecuniary advantages from their formation.

Another objection alleged against canals, on public grounds, was alike speedily disproved. It was said that inland navigation, by reason of its greater cheapness, ease, and certainty, must necessarily diminish the coasting trade, and consequently discourage the training of seamen, who formed the constitutional bulwark of the kingdom. But the extraordinarily rapid growth of the shipping-trade of Liverpool, and the vastly increased number of sea-going vessels required to accommodate the traffic converging on that seaport, very soon showed that canals, instead of diminishing, were calculated immensely to promote the naval power and resources of England. Thus it was found that in the thirty years which elapsed subsequent to the opening of the Duke's Canal between Worsley and Manchester,—during which time the navigation had also been opened to the Mersey, and the Grand Trunk and other main canals had been constructed, connecting the principal inland towns with the sea-ports,—the tonnage of English ships had increased threefold, and the number of sailors been more than doubled.

So great an impulse had thus been given to the industry of the country, and it had become so clear that facility of communication must be an almost unmixed good, that a desire for the extension of canals sprang up in all districts; and instead of being resisted and denounced, they became everywhere the rage. They were advocated in pamphlets, in newspapers, and at public meetings. One enthusiastic pamphleteer, advocating the formation of a canal between Kendal and Manchester, denounced the wretched state of the turnpike-roads, which were maintained by "an enormous tax," and exclaimed, "May we all scorn to plod through the dirt as we long have done at so large an expense; and for the support of our drooping manufactories, let canals be made through the whole nation as common as the public highways." *

There seemed, indeed, to be every probability that this desire would be shortly fulfilled; for so soon as the canals which had been made began to pay dividends, the strong motive of personal gain became superadded to that of public utility. The rapid increase of wealth which they promoted served to stimulate the projection of new schemes; and in a very few years after Brindley's death we find an immense number of Navigation Acts receiving the sanction of the legislature, and canal works in progress in all parts of the country. The shares were quoted upon 'Change, where they became the subject of commerce, and very shortly of wild speculation.

By the year 1792, the country was in a perfect ferment about canal shares. Notices of eighteen new canals were published in the 'Gazette' of the 18th August in that year. The current premiums on single shares in those companies for which Acts had been obtained were as follows: Grand Trunk, 350*l.*; Birmingham and Fazeley, 1170*l.*; Coventry,

* 'A Cursory View of a proposed Canal from Kendal to the Duke of Bridgewater's Canal, leading to the great manufacturing town of Manchester.' 1785.

350*l.*; Leicester, 155*l.*; and so on. There was a rush to secure allotments in the new schemes, and the requisite capitals were at once eagerly subscribed. At the first meeting, held in 1790, of the promoters of the Ellesmere Canal, 112 miles in extent, to connect the Mersey, the Dee, and the Severn, applications were made for four times the disposable number of shares.

A great number of worthless and merely speculative schemes were thus set on foot, which brought ruin upon many, and led to waste both of labour and capital. But numerous sound projects were at the same time launched, and an extraordinary stimulus was given to the prosecution of measures, too long delayed, for effectually opening up the communications of the country. The movement extended to Scotland, where the Forth and Clyde Canal, and the Crinan Canal, were projected; and to Ireland, where the Grand Canal and Royal Canal were undertaken. But, as Arthur Young pithily remarked, in reference to these latter projects, "a history of public works in Ireland would be a history of jobs."

In the course of the four years ending in 1794, not fewer than eighty-one Canal and Navigation Acts were obtained: of these, forty-five were passed in the two latter years, authorising the expenditure of not less than 5,300,000*l.* As in the case of the railways at a subsequent period, works which might, without pressure upon the national resources, easily have been executed if spread over a longer period, were undertaken all at once; and the usual consequences followed, of panic, depreciation, and loss.

But though individuals lost, the public were

eventually the gainers. Many projects fell through, but the greater number were commenced, and after passing through the usual financial difficulties, were finished and used for traffic. The country became thoroughly opened up in all directions by about 2600 miles of navigable canals in England, 276 miles in Ireland, and 225 miles in Scotland. The cost of executing these great water-ways is estimated to have amounted to about fifty millions sterling. There was not a place in England south of Durham, more than fifteen miles from water communication ; and most of the large towns, especially in the manufacturing districts, were directly accommodated with the means of easy transport of their goods to the principal markets. "At the beginning of the present century," says Dr. Aiken, writing in 1795, "it was thought a most arduous task to make a high road practicable for carriages over the hills and moors which separate Yorkshire from Lancashire, and now they are pierced through by three navigable canals ! "

Notwithstanding the great additional facilities for conveyance of merchandise which have been provided of late years by the construction of railways, a very large proportion of the heavy carrying trade of the country still continues to be conducted upon canals. It was indeed at one time proposed, during the railway mania, and that by a somewhat shrewd engineer, to fill up the canals and make railways of them. It was even predicted, during the construction of the Liverpool and Manchester Railway, that "within twelve months of its opening, the Bridgewater Canal would be closed, and the place of its waters be covered over with rushes." But canals have stood their ground, even against

railways; and the Duke's Canal, instead of being
closed, continues to carry as much traffic as ever.
It has lost the conveyance of passengers by the
fly-boats,* it is true; but it has retained and in
many respects increased its traffic in minerals and
merchandise. The canals have stood the competi-
tion of railways far more successfully than the old
turnpike-roads, though these too are still, in their
way, as indispensable as canals and railways themselves. Not less than twenty millions of tons of
traffic are estimated to be carried annually upon
the canals of England alone, and this quantity is
steadily increasing. In 1835, before the opening of
the London and Birmingham Railway, the through
tonnage carried on the Grand Junction Canal was
310,475 tons; and in 1845, after the railway had
been open for ten years, the tonnage carried on the
canal had increased to 480,626 tons. At a meeting
of proprietors of the Birmingham Canal Navigations, held in October, 1860, the chairman said,
"the receipts for the last six months were, with
one exception, the largest they had ever had."

Railways are a great invention, but in their
day canals were as highly valued, and indeed quite
as important; and it is fitting that the men by

---

* The following curious paragraph is from the 'Times' of the
19th December, 1806. It relates to the despatching of troops from
London for Ireland, during a time of great excitement :—"The
first division of the troops that are to proceed by Paddington
Canal for Liverpool, and thence by transports for Dublin, will
leave Paddington to-day, and will be followed by others to-morrow,
and on Sunday. By this mode of conveyance the men will be·
only seven days in reaching Liverpool, and with comparatively
little fatigue, and it would take them above fourteen days to march
that distance. Relays of fresh horses for the canal-boats have
been ordered to be in readiness at all the stations."

whom they were constructed should not be for-
gotten. We may be apt to think lightly of the
merits and achievements of the early engineers,
now that works of so much greater magnitude are
accomplished without difficulty. The appliances of
modern mechanics enable men of this day to dwarf
by comparison the achievements of their prede-
cessors, who had obstructions to encounter which
modern engineers know nothing of. The genius
of the older men now seems slow, although they
were the wonder of their own age. The canal, and
its barges tugged along by horses, may appear a
cumbersome mode of communication, beside the
railway and the locomotive with its power and
speed. Yet canals still are, and will long continue
to form, an essential part of our great system of
commercial communication,—as much so as roads,
railways, or the ocean itself.

Brindley's Burial-place at New Chapel.

APPENDIX.

THE GRAND CANAL OF LANGUEDOC AND ITS CONSTRUCTOR, PIERRE-PAUL RIQUET DE BONREPOS.

THE Canal du Midi, more commonly known as the Grand Canal of Languedoc, was one of the most important works of the kind at the time at which it was executed, though it has since been surpassed by many canals in France, as well as in England and other countries. It was commenced in 1666, about a hundred years before Brindley began the Bridgewater Canal, and it was finished in 1681. The magnitude and importance of the work will be understood from the following brief statement of facts.

The canal crosses the isthmus which connects France with Spain, along the valley between the Pyrenees and the Rhone, and extends from the Garonne at Toulouse (which is navigable from thence to the Bay of Biscay) to Cette on the shores of the Mediterranean, thus uniting that sea with the Atlantic Ocean. The length of the navigation, from Toulouse to Cette, is about 158 English miles, including its passage through Lake Thau, near Cette, where, in consequence of the shallowness of the lake, it is confined for a considerable distance within artificial dykes. The canal is carried along its course over rivers, and under hills, by means of numerous aqueducts, bridges, and tunnels. From the Garonne to the summit it rises 207 feet by twenty-six locks ; the summit level is 3½ miles, after which it descends by thirty-seven locks into the Aude near Carcassonne. It then proceeds along the north side of that river, passing over several streams, and descending by twenty-two locks into Lake Thau. There are other locks in the neighbourhood of Toulouse and Cette ; the whole number being above a hundred. The fall from the summit at Naurouse to the Mediterranean is 621½ feet ; a fact of itself which bespeaks the formidable character of the undertaking.

The Grand Canal of Languedoc was constructed by Pierre-Paul Riquet de Bonrepos, a man of extraordinary force of character, bold yet prudent, enterprising and at the same time sagacious and patient, possessed by an inexhaustible capacity for work, and endowed with a faculty for business, as displayed in his organisation of the labours of others, amounting to genius. Yet Riquet, like Brindley, was for the most part self-taught, and was impelled by his instincts rather than by his education to enter upon the construction of canals, which eventually became the great business of his life. Presuming that an account of the " French Brindley " will not be uninteresting to English readers, we append the following summary of Riquet's career in connection with the great enterprise in question.

The union of the Mediterranean with the Atlantic by means of a navigable canal across the South of France had long formed the subject of much curious speculation. The design of such a work will be found clearly sketched out in the ' Mémoires de Sully ; ' but the project seemed to be so difficult of execution, that no steps were taken to carry it into effect until it was vigorously taken in hand by Riquet in the reign of Louis XIV. Though descended from a noble stock (the Arrighetti or Riquetti of Florence, to a branch of which Riquetti Marquis de Mirabeau belonged), Riquet, at the time he took the scheme in hand, was only a simple exciseman (homme de gabelle). His place of residence was at the village of Bonrepos, situated near the foot of the Montagne Noire, where he owned some property.

France is there at about its narrowest part, and it had naturally occurred to those who speculated on the subject of a canal, that it would be of great public importance if by such means the large navigable river, the Garonne, which flowed into the Western Ocean, could be united to the smaller river, the Aude, which flowed into the Mediterranean. Both had their sources in the Pyrenees, and in one part of their course the rivers were only about fourteen leagues apart. The idea of joining them was thus perfectly simple. The great difficulty was in the execution of the work, the levels being different, and the intervening country rocky and mountainous. The deputies from Languedoc to the States General of Paris had at various times brought the subject of the proposed canal under the notice of the Government, and engineers were even sent into the locality to report as to the feasibility of the scheme ; but the result of their examination only served to confirm the general impression which prevailed as to the impracticable character of the undertaking.

The situation of Riquet's property at the Montagne Noire

probably had the effect of directing his attention to the subject of the proposed canal. Though the king had made him a tax-gatherer, nature had made him a mathematician; and his studies having taken a practical turn, he gradually went from geometry to levelling and surveying. His instruments and appliances were of the simplest sort, but they proved sufficient for his purpose. The Chancellor d'Aguesseau, in the memoir of his father, who personally knew Riquet, says of him that "the only instrument he then possessed was a wretched compass of iron; and it was this, with very little instruction and assistance, that led him, guided mainly by his powerful natural instinct, which often avails more than science, to form the daring conception of uniting the Ocean with the Mediterranean." *

He carefully examined the course of the streams in his neighbourhood, trying to find out some practicable method of uniting the Garonne and the Aude, but more especially some means of supplying the upper levels of the Canal which he had in his mind's eye with sufficient water for purposes of navigation. With those objects he made many surveys of the adjacent country, until he knew almost every foot of the ground for thirty miles round. In the mean time he hired one Pierre, the son of a well-sinker of Revel, to dig a number of experimental little canals under his direction, in his gardens at Bonrepos, where they are still to be seen. These miniature works included conduits, sluices, locks, and even a model tunnel through a hill.

At length, in the year 1662, he brought his plans under the notice of the famous Minister Colbert. Addressing him from the village of Bonrepos, Riquet said, "You will be surprised that I should address you on a matter about which I might be supposed to know nothing, and that an exciseman should mix himself up with river surveying. But you will excuse my presumption when I inform you that it is by the order of the Archbishop of Toulouse that I write to you." † He then proceeded to state that, having made a particular study of the subject, he had formed definite plans for carrying into effect the proposed canal, of which he enclosed a description, though in a very imperfect form; "for," he added, "not having learnt Greek nor Latin, and scarce knowing how to speak French, it is not possible for me to explain myself without stammering."

* 'Œuvres de d'Aguesseau,' tome xiii.
† 'Histoire du Canal de Languedoc,' par les Descendants de Pierre-Pau Riquet de Bonrepos. Paris, 1805.

Pointing out the great advantages to the nation of the proposed canal, the time and the money that would be saved by enabling French ships to avoid the long voyage between the west coast and the Mediterranean by the Straits of Gibraltar, while the resources of the rich districts of Languedoc and Guienne would be freely opened up to the operations of commerce, Riquet concluded by stating that when he had the pleasure of learning that the project met with the general approval of the Minister, he would send him the details of his plans, the number of locks it would be necessary to provide, together with his calculations of the exact length, breadth, depth, and other particulars of the proposed canal.

Colbert, at that time Controller-General of Finance, was actively engaged in opening up new sources of wealth to France, and Riquet's plan at once attracted his attention and excited his admiration. He lost no time in bringing it under the notice of Louis XIV., whose mind was impressed by all undertakings which bore upon them the stamp of greatness. The king saw that Riquet's enterprise, if carried out, was calculated to add to the glory of his reign ; and he resolved to assist it by all the means in his power. By his order, a Royal Commission was appointed to inquire into the scheme, examine on the spot the direction of the proposed navigation, and report the result.

Meanwhile Riquet was not idle. He walked over the entire line of the intended canal several times, correcting, amending, and perfecting the details of his plans with every possible care. "I have gone everywhere, all over the ground," he wrote to the Archbishop, "with level, compass, and measuring line, so that I am perfectly acquainted with the route, the various lengths, the locks that will be required, the nature of the soil, whether rock or pasture, the elevations, and the number of mills along the line of navigation. In a word, my Lord, I am ignorant of no point of detail in the project, and the plan which I am prepared to submit will be faithful, being made on the ground, and with full knowledge of the subject." His survey finished, Riquet proceeded to Paris to see Colbert, to whom he was introduced by his friend the Archbishop ; and after many conferences, Riquet returned to Languedoc to prepare for the inspection of the proposed line of navigation by the Commission. Their labours extended over two months, beginning at Toulouse and ending at Beziers.*

* The Procès-Verbal des Commissaires is preserved in the Manuscrits Colbert, No. 202, Bibliothèque Impériale.

The result of the investigation was favourable to Riquet's plan, which was pronounced practicable, with certain modifications, more particularly as respected the water-supply ; and the Commissioners further recommended the extension of the proposed canal to a harbour at Cette.

A long correspondence ensued between Riquet and Colbert as to the details of the scheme. Riquet disputed the conclusions of the Commissioners as to the alleged difficulty of constructing the works near Pierre de Naurouse intended for the supply of water to the canal ; and, to show his confidence in his own plans, he boldly offered to construct the conduit there at his own risk and cost : "in undertaking to do which," said he, "I risk both honour and goods ; for if I fail in satisfactorily performing the work, I shall pass for a mere visionary, and at the same time lose a considerable sum of money." So much difference of opinion prevailed upon this point,—the men of science alleging the inadequacy of Riquet's plans, and Riquet himself urging their perfect sufficiency,—that Colbert resolved that until this point was decided, no further step should be taken to authorise the canal itself to be proceeded with.

But in order that Riquet might have an opportunity of vindicating the soundness of his plans, and not improbably with a view to test his practical ability to carry out the larger works of excavation and construction, letters patent were issued entrusting him to proceed with the necessary trench or conduit to enable some experience to be obtained of the inclination of the ground and the probability or otherwise of obtaining an adequate supply of water for the navigation.

Riquet, with his usual promptitude, immediately began the execution of this work. The rapidity with which he proceeded surprised everybody ; and the conduit was very speedily finished to the entire satisfaction of the Government inspectors. Having given this proof of his practical skill, and demonstrated the possibility of supplying sufficient water to the summit level, the king determined on authorising Riquet to proceed with the construction of the canal.

The question of providing the requisite money for the purpose formed the next important subject of consideration. As the province of Languedoc would derive the principal advantages from the navigation being opened out, it was proposed at the assembly of the States, in 1665, that that province should contribute a certain proportion of the cost, on condition that the king should provide the remainder. Notwithstanding, however, the great advantages to the province of the construction of the canal, the States of Languedoc would not untie their purse-strings ; and they declared (26th February, 1666) that they would neither then nor thereafter contribute towards its cost. On the other hand, the royal treasury had become almost exhausted, and any new expense could with difficulty be borne by it.

Riquet's demonstration of the possibility of uniting the Mediterranean with the ocean had therefore thus far been in vain. But he had fairly committed himself to the enterprise, and he was not the man now to turn back. Having had the courage to conceive the design, he urged the Government to allow the work to proceed ; and he suggested a method by which in his opinion the means might be provided without unnecessarily burdening the public treasury. He offered to construct the works upon the first division of the canal, from Toulouse to Trebes near the river Aude, within a period of eight years, for the sum of 3,630,000 livres ; and in payment of this sum he proposed to the king that he should grant him the sole farming of the taxes in Languedoc, Roussillon, Conflans, and Cerdagne during six years, on the same terms at which they were then held, and also that the offices of Controllers of peasants'-tax (des tailles), of tradesmen's-tax, and of rights over the salt-works of Pecais, should at the same time be assigned to him. This was a bold offer ; but the Council of State accepted it, and the necessary legal powers were accordingly granted to Riquet, upon which measures were taken to enable him to proceed with the works.

The plans which Riquet had prepared were revised and settled by M. Cavalier, the Commissaire-Général of Fortifications, who acted on behalf of the king. Riquet's plans had been prepared with such care, that in his specification Cavalier for the most part adopted them, only altering the dimensions,* which were somewhat increased. The King's Engineer was careful not to define too rigidly the manner in which the works were to be carried out. He did not even trace a definite line to be followed, but merely marked the general limits within which the canal was to run. He foresaw that with experience, many modifications might be found necessary in the progress of the works, and he preferred leaving the management of the details as much as possible to the skill and judgment of the contracting engineer.†

* ' Canaux Navigables,' par La Lande, 120.
† ' Histoire du Canal de Languedoc,' p. 50.

Riquet had now to display his genius in a new sphere. Hitherto he had exhibited himself mainly as a designer; he had drawn plans, advocated and explained them to others, and transacted the part of a diplomatist in getting them adopted. Though his execution of the conduit near Pierre de Naurouse had enabled him to give satisfactory proofs of his engineering skill, the work he was now about to enter upon was of a much more formidable character, calling for the exercise of a varied class of practical qualities. He had to direct the labours of a very large number of men, to select the most suitable persons to superintend their various operations, and meanwhile to give his continuous attention to the carrying out of his plans, which as in the case of all great undertakings, required constant modification according to the many unforeseen circumstances which from time to time occurred in the course of their execution.

Anxious to proceed with the greatest despatch, Riquet began with the organisation of his staff of workmen and superintendents. He divided them into a number of distinct groups, appointing a chef d'attelier to each, under whom were five brigadiers, each brigadier having the direction of fifty workmen. These groups were again combined in departments, a controller-general being appointed over each, and under him were travelling controllers, who received the reports of the brigadiers and chefs d'atteliers, and thus effectually maintained and concentrated the operations of the workpeople, who sometimes numbered from eleven to twelve thousand.

Before Riquet had proceeded far with his enterprise, he experienced a difficulty which has proved the bane of many a grand scheme—want of money! The produce of the taxes above referred to was not sufficient to enable him to push on the works with vigour; but, rather than they should be delayed, he himself incurred heavy debts, selling or mortgaging all his available property to raise the requisite means. Notwithstanding the early decision of the States of Languedoc not to contribute towards the cost of constructing the canal, Riquet again and again made urgent appeals to them for help, but for some time in vain. Several grants had been made from the royal treasury to enable the contract to proceed, but Louis XIV. having become engaged in one of his expensive wars, was no longer able to contribute; and Riquet, having come to the end of his own resources, began to fear lest the canal works should be brought to a complete standstill.

Colbert continued his fast friend and supporter, and took the most lively interest in the prosecution of the undertaking. His name was indeed a tower of strength, and his influence was in itself equal to a large capital. Of this Riquet on one occasion made very adroit use, for the purpose of inducing the States of Languedoc at length to take part in his enterprise. It is said that, in order to impress upon their minds the confidence reposed in him by the great Minister, and thus to enhance his credit with them, he persuaded Colbert to allow him to try the following ruse. He asked to be permitted to enter his cabinet while he was engaged with the farmers-general of the province in discussing the renewal of their leases. Colbert consented ; and while thus occupied, Riquet turned the key of the room-door, and entered and sat down, without saying a word to any one, and without any one speaking to him. The farmers-general looked at Riquet, then at the Minister, who took no heed of him, and then at each other. Strange that Colbert should place such confidence in Riquet as to permit him thus to enter his private chamber at pleasure ! A second meeting of the same kind took place, and again Riquet entered as before. After the interview was over, the farmers spoke to Riquet of his canal, and its great utility ; and ended by offering to lend him 200,000 livres. Riquet, however, listened to the proposal very coolly, without accepting it. At the end of a third interview with the Minister, at which Riquet was present as before, the farmers raised their terms, and offered to lend him 500,000 livres. Riquet replied that he could do nothing without the sanction of the Minister ; and re-entering the cabinet of Colbert, he related to him what had passed. The Minister was very much amused at Riquet's adroitness, and readily give his sanction to the proposed loan.*

This advance proved the beginning of a series of loans of great magnitude advanced to Riquet by the States of Languedoc to enable him to complete the canal. Though they were slow to believe in the practicability of the scheme, and for some time regarded it as impossible, their views changed when they saw the first part of the canal completed from Toulouse to Trebes, and opened for the purposes of navigation. They were then ready to recognise its great public uses, and from that time forward they exerted themselves to raise the necessary money to enable it to be completed to its junction with the Mediterranean at the port of Cette.

* This anecdote is given in the ' Mémoires du Baron de Bésenval.'

But Riquet had numerous difficulties to encounter besides those arising from want of money in the course of his undertaking. The prosecution of the works involved constant anxiety and unremitting labour. One of his greatest troubles was the conciliation of the owners of the land through which the canal passed, many of whom were extremely hostile to the enterprise, and feared that it would inflict irreparable injury upon their property. Worse than all was the malice, misrepresentation, and calumny that pursued him. He was denounced as an impostor, attempting to do an impossible thing ; he was wasting public money upon a work which, even if finished, could never be of any use. Before he began the canal, it was predicted that water enough could never be collected to supply the summit level ; and after that difficulty had been satisfactorily solved, local detraction was directed against the works. " Indeed," writes M. de Froidour to M. de Barillon, " the people of the locality are so agreed in decrying them that the wonder is to find a person who has not arrived at the foregone conclusion that this enterprise can never succeed." *

Notwithstanding the alleged uselessness of the canal, the insufficiency of the works, and the prophecies of its failure even if completed, Riquet bravely bore up, amidst toil, and disappointments, and bodily suffering. He never lost hope or courage, but persevered through all. Writing to Colbert in April, 1667, he said, " I now know the strong as well as the weak points of my work better than I before knew them ; and I can assure you in all truth that it would not be easy to imagine a grander or more useful undertaking ; I have all the water-supply that I require, and the invention of my reservoirs will furnish me during summer with sufficient to render the navigation perpetual."

At another time he wrote to Colbert :—" My enterprise is the dearest of my offspring : I look chiefly to the glory of it, and your satisfaction, not my profit ; for though I wish to leave the honour to my children, I have no ambition to leave them great wealth." And again—" My object is not to enrich myself, but to accomplish a useful work, and prove the soundness of my design, which most people have hitherto regarded as impossible."

About the beginning of 1670, after three years' labour,

* Three Letters from M. de Froidour to M. de Barillon, published at Toulouse in 1672.

I. 2 B

part of the canal was opened, from Toulouse to Duperier, and used for the transport of materials. This was a comparatively easy section of the undertaking. But Riquet was desirous of exhibiting the practical uses of the canal at the earliest possible period, not only for the purpose of mitigating the popular opposition, but of encouraging the king, Colbert, and the States-General, to support him with the necessary means to complete the remainder of the navigation from Trebes to Cette. Two years later, a further portion was finished and opened for public traffic. The Archbishop embarked at Naurouse, and sailed along the new canal down to Toulouse. Four large barques ascended from the Garonne to Naurouse, and returned freighted with provisions and merchandise. The merchants of Gaillac, who had before been unable to send their wines to Bordeaux for sale, were now enabled to do so, and they established a packet-boat which regularly went between Naurouse and Toulouse three times a week.

The remaining portions of the canal were in full progress. The basins, the conduits, the locks, were all well advanced as far as Castelnaudary, and Riquet was vigorously grappling with and successively overcoming the great difficulties which occurred in the construction of the works between that place and the Mediterranean. Not the least among the number of his obstructions were the quarrels between the two Commissioners appointed by the king on the one hand, and by the States of Languedoc on the other, to superintend the execution of the undertaking. Each represented particular local interests, and whilst one desired to keep the canal to the north, the other wished it to proceed more to the south by way of Narbonne. Between their contentions, Riquet had sometimes a difficult course to steer. Thus, at Malpas, where it was necessary to carry the canal by a tunnel under the hill of Enserune, both Commissioners pronounced the work to be impracticable, because the hill appeared to consist chiefly of a sandy stuff permeable to water, and apt to give way. But they were far from agreed as to the remedy, each urging upon Riquet the adoption of an opposite course, one that he should carry the canal northward by Maureillan, the other that he should carry it southward by Nissau and Vendres. They both wrote to Colbert, pointing out that Riquet's plan could never be carried out; that the scheme threatened to prove a total failure, because his work had run its head into a sand-hill, at a point where it had a lake on each side of it, from twenty-five to thirty feet below its level. In the mean time the Commissioners gave Riquet orders to suspend the further prosecution of the works. Riquet put the orders in his pocket, and coolly resolved to carry out his own plan. To conceal his design, he pretended to abandon the trench leading to the mountain, and sent the workmen to that part of the canal which lay between Beziers and Agde. He then privily set a number of excavators to work upon the mountain-side near Malpas, and in six days he had vanquished "impossibility," and cut a clear passage through it for his canal! When the work was finished, he sent invitations to the Cardinal de Bonzy and to the two Commissioners to come and inspect what he had done. Greatly to their surprise, he led them right through his tunnel, lit up with flambeaux; and his triumph was complete.

It was not so easy to overcome the constantly recurring difficulties occasioned by the want of money. Nothing but money would satisfy his thousands of workmen and work-women (for of the latter about six hundred were employed), and he was often put to the greatest straits for want of it. We find him, in 1675, writing to Colbert in very urgent terms. Unless supplied with funds, he represented that it would be impossible for him to go on. " People tell me," said he, "that I am only digging a canal in which to drown myself and my family." The king ordered a remittance to be sent to Riquet, but it was insufficient for his purposes. The costly harbour works at Cette were now in full progress, absorbing a great deal of money; and Riquet's creditors grew more and more clamorous. Colbert urged the States of Languedoc to give him more substantial help, and they complied so far as to vote him a loan of 300,000 livres. But this money was only to be advanced at different and remote periods. Thus it did little to help him either in credit or in funds. " I am doing everything that is possible," he again wrote to Colbert, on the 21st of January, 1679, "to find persons that will lend me money to enable me to finish the canal in the course of the present year. But I am so overwhelmed with debt, that nobody will trust me, so that I am under the necessity of again having recourse to you, and informing you of my needs; you will see what they are by the enclosed statement; I venture to ask that you will state your wishes alongside of each item, so that I may be put in a position to bring my enterprise to a successful termination : it is my passion, and I shall be in despair if I cannot finish it. Time flies; and once lost, it can never be recovered."

2 B 2

Riquet had indeed reason to be apprehensive that he might not live to complete his work. The strain upon his mind and body for the fifteen years during which the canal was in progress had been very great, and he had more than one serious attack of illness. But still the enterprise went forward. The organisation established by him was so perfect that his occasional absence from this cause was scarcely felt; besides, his eldest son was now in a manner competent to supply his place.

The third portion of the undertaking, consisting of the harbour and sea entrance to the canal at Cette, was being vigorously pushed on, and the canal was almost ready for opening throughout from one end of it to the other, when, worn out by toil and disease, Riquet breathed his last, without having the satisfaction of seeing his glorious work brought to completion. The canal was finished under the superintendence of his son, and was opened for public traffic about six months after Riquet's death.

The total cost of the canal was about sixteen million livres, equal to 1,320,000*l*. sterling. Riquet sank all his own means in it, and when he died it was found that his debts amounted to more than two million livres. To defray them, his representatives sold the principal part of the property he held in the concern, and it was not until the year 1724, forty years after the navigation had been opened, that it began to yield a revenue to Riquet's heirs.

As in the case of many other great works, Riquet's merits have not been left undisputed. Thirty years after his death, a claim was set up on behalf of one of his assistants, M. Andreossy, as having been the designer of the canal; while, more recently, a like claim has been made on behalf of M. Cavalier, the king's engineer. Although all contemporary witnesses had died before Riquet's merits, till then undisputed, were thus called in question, happily the Archives of the canal as well as the Colbert correspondence survive to prove that, beyond the shadow of a doubt, Riquet was not only the inventor and the designer, but the constructor, of the Grand Canal of Languedoc.

It must, however, be admitted that the genius of this great engineer has received but slight recognition among his own countrymen, there not being a single tolerable memoir of Riquet to be found in the whole range of French literature. To this, however, it might be rejoined, that only a few years since the same thing might have been said of our own Brindley.

INDEX.

VERMUYDEN.

the works destroyed, 33 ; would drain Cambridge Fens ; contracts to purchase Malvern Chase, 35 ; takes land on Sedgemoor, and leases a lead-mine ; legal troubles ; imprisoned ; engaged with the drainage of the Great Level of the Fens ; offers to find funds, and carry out the works ; contracts, 35-6 ; contract abrogated ; starts again with the work under Earl of Bedford ; money difficulty ; his energy, 38 ; the principal works, 38-9 ; opposition of the Fen-men, 39-40 ; defects of his system, 44 ; King Charles undertakes the work, and employs Vermuyden ; his new design, 44-5 ; Cromwell's opposition, 45-6 ; the works destroyed ; he remains undaunted, 48 ; he again resumes the work, 49 ; employs Scotch and Dutch prisoners ; the new works, 50 ; the works completed, 52 ; all his property swallowed up ; dies in poverty and obscurity, 54 ; works on the Thames, 126-7. Vermuyden, Cornelius, jun., Colonel, 48.
Vernatti, F., Dutch capitalist, 26, 30, 32.

WAGES of workmen, 253.
Waldersea County Drainage Act, 19.
Walker, author of ' The Original,' 197.
—— Colliery, Newcastle, engine at, 185.
Wallend Marsh, 4-5.
Walloon immigrants, 27.
Walpole, Horace, letters, 188, 191-2.
Wapping Marsh, or High Street, 20.
Ware, New River Head, 82,
Warrington, Lancashire, 151, 187, 201, 224, 228-44.

YOUNG.

Wash, the Great, 10-12.
Water as an agent, and in engineering, 58-64.
Water-carriers of old London, 95.
Weaver Navigation, 152, 186, 302, 311, 316.
Wedgwood, Josiah ; his character ; early life ; infirmity, 306 ; commences business ; Ivy House, 308-9 ; extension of his trade, 309 ; promotes the Grand Trunk Canal scheme, 311, 313 ; cuts the first sod, 319 ; evidence as to advantages of canal, 328-9, 334.
Welland River, in the Fens, 10, 50.
Wells of old London, 58-61.
Wesley, John, at Burslem, 304-5, 330.
Westerdyke, Dutch engineer, 38, 49.
Whittlesea Mere, 14, 39, 51, 56.
Whitworth, Sir R., 303, 315, 353.
——, Brindley's pupil, 334.
Wiches, the Cheshire, 311, 316.
Wigan, canal to, 186.
Wight, Isle of, 100.
Wilden, Derby, 300, 314, 318, 321, 326.
Winchelsea, Romney Marsh, 5.
Windsor Royal Park drained by Vermuyden, 22.
Wisbeach, 27, 39, 45, 51.
Witham breach repaired, 20.
—— River, in the fens, 10, 12 ; floated ships to Lincoln, 13.
Wolverhampton Canal, 332, 336.
Worsley Basin, 215-17.
—— Mill, 202, 209, 215, 231-2.
—— Old Hall, 207, 265-7.
Wren, Sir Christopher, low remuneration of, 175.

YARMOUTH Haven constructed, 20.
Yarranton, Andrew, his projects, 149.
Young, Arthur, extracts from, 215, 219, 220, 247, 305.

THE END

WILLIAM CLOWES AND SONS, LIMITED, LONDON AND BECCLES.